To Marilyn for her
infinite patience
and encouragement

A large portion of this was previously published as part of *Barron's How to Prepare for High School Equivalency Examinations—Beginning Preparation in Mathematics*, © 1977.

All inquiries should be addressed to:
Barron's Educational Series, Inc.
250 Wireless Boulevard
Hauppauge, New York 11788

Library of Congress Catalog Card No. 78-21490

International Standard Book No. 0-8120-691-7

Library of Congress Cataloging in Publication Data
Williams, Edward, 1926-
Mathematics.
(Barron's back to basics series)
CONTENTS: 1. From addition to division. 2. Improving your skills with fractions. 3. Percents and other matters.
1. Mathematics — 1961- [1. Mathematics]
I. Title.
QA39.2.W5583 510 78-21490
ISBN 0-8120-0691-7

PRINTED IN THE UNITED STATES OF AMERICA

456 100 12 11 10

Contents

In this series we plan to help students improve their skills in mathematics. You know from daily living that mathematics must be used in many ways. Basic arithmetic — adding, subtracting, multiplying, and dividing — applies to topics such as buying and selling, wages, banking, insurance, taxes, and measurement. As an adult you will have to deal with these problems and you will need more advanced skills.

In this series of books we will provide the explanations and practice for these mathematical skills. We also show you how to study topics of mathematics in a logical and practical order. You should try the test at the beginning of each chapter to see if you understand the ideas well or need to review them. The series is set up so you can move at your own pace.

Barron's Back to Basics Series — Mathematics — will help you advance to high school level work. Topics near the end of the third volume will be more advanced, but you will be ready for them. Many others have used these skills to move ahead to better grades and an improved understanding of mathematics. Your plans for yourself are important; we want to help you make them come true.

I would like to acknowledge the assistance of the many people who helped to shape this series. My thanks go to Dr. Eugene Farley, to Lester Schlumpf, and to Linda Bucholtz Ross.

introduction

Barron's Back to Basics Series — Mathematics reviews and explains basic mathematics. For those of you who want to improve your chances for a better job or to be better informed when you buy something or do business with someone, mathematical skills are important. With this series, you can move from beginning level mathematics to high school level, at your own speed. You can use this series on your own or in class with a teacher to help.

What This Book Can Do

Barron's Back to Basics Series — Mathematics will try to:

1. Review and explain the basic ideas and skills of beginning mathematics.
2. Give many examples to practice the ideas and skills.
3. Increase your confidence to try more advanced material.
4. Encourage you to do better on your mathematics tests in school.

You could summarize the purposes of this book as *building your skills and confidence*.

How The Series Is Organized

We emphasize basic mathematics — addition, subtraction, multiplication, and division. Each of these is applied to whole numbers, fractions, decimals, and measurement. Later books introduce topics you will cover more at high school level. The ideas and skills are related to problems of everyday life at work or at home.

Each chapter begins with a test so that you can check whether you know and understand the ideas. This is called a *pretest* because it comes *before* the chapter work. If you do well on the pretest you can save time by skipping the material in that chapter. If you find you need some help, you have the advantage of knowing which ideas are not clear to you. Then you can read the chapter with a specific purpose. Each topic and each method will be carefully explained. Practice examples are given to help you learn the idea.

After studying each chapter you can check how well you have learned by taking the *posttest*, so named because it comes *after* the chapter work. If you do well on the posttest, you can feel confident about moving to the next chapter. You may find some problems that you cannot do. To help you, there will be a section at the end of the chapter which guides you to the parts you need to study again.

The standard for doing well varies for each test. You can progress as fast as you can study and understand. Answers are given so that you may check whether you did the problems correctly. Obviously, there is no value to looking at the answer unless you have worked the problem. You want to know *how* to do the problems and to learn the technique so that you can use it on similar problems.

For additional interest and a little relaxation, there are some mathematical games and puzzles scattered throughout the book. These also give you practice in the basic mathematical skills.

How To Use This Book

Each individual studies in his or her own way. However, the series is planned in such a way that the following suggestions should work best for you:

1. Read the brief introduction to the topic at the beginning of a chapter.
2. Take the pre-test. Check your answers to decide whether
   a. You need to study that chapter, or
   b. You can move to the next chapter.
3. If you are not sure, study the material in the chapter by
   a. Reading the explanations and
   b. Working the practice examples.
   (Note – Practice examples are necessary to gain good control over the material. Do not skip the practice until you are sure you do the problems well.)
4. Take the posttest at the end of the chapter to see how well you have learned.
5. Stress explanations and methods.
   *(Note – You are urged to spend most of your time on **explanations.** The **how** and **why** of mathematics, or the **methods** and the **reasons,** are the most important parts. Once you understand the **reason** for something and the **method** or way to do it, you can apply the knowledge to many other problems.)*

Why This Is Important

Only you can decide how important education and skills are to you. Total lifetime earnings can be increased by getting more training. This is called the *economic* value of education. There is no question that better jobs, chances for promotion, and new kinds of training are related to learning skills. Employers try to select workers who gain new skills easily and rapidly. Advancement or promotion on your present job or to a different kind of job often depends upon the skills that will be covered in this book.

Personal values cannot be figured in terms of money or job, but they are of great importance. You gain in self-respect and confidence when you know that your skills have improved. Perhaps you just want to prove to yourself or others that you can be better. There is no point in regretting your past failures to achieve. There may be good reasons why you did not gain skill in your previous education. What counts now are the changes you want to make. If you want to concentrate on the past, this book cannot help you. If you want to think of the present and your future, help is available.

Perhaps you have been away from learning for a while. You may even feel that it is too much effort to study again. Let us not talk about reasons that keep you from trying to do your best. The important action now is to apply your energy to renewing habits and skills. Each skill you learn helps you to learn more. The habit of study leads to success and your success will encourage you to try further. You have already shown faith in yourself by reading this book.

Let's Get Started

It is a good idea to talk about the future. We can benefit from setting a goal or planning ahead. However, our concern for now is to get started on a review of mathematics. Chapter 1 starts the refresher course by discussing numbers and number systems.

reference tables

United States System Of Weights And Measures

Units of Length

12 inches (in.) = 1 foot (ft.)
3 feet (ft.) = 1 yard (yd.)
1,760 yards (yd.) = 1 mile (mi.)

Units of Capacity

16 fluid ounces (fl. oz.) = 1 pint (pt.)
2 pints (pt.) = 1 quart (qt.)
4 quarts (qt.) = 1 gallon (gal.)

Units of Weight

16 ounces (oz.) = 1 pound (lb.)
2,000 pounds (lb.) = 1 ton (T.)

Units of Time

60 seconds (sec.) = 1 minute (min.)
60 minutes (min.) = 1 hour (hr.)
24 hours (hr.) = 1 day (da.)
7 days (da.) = 1 week (wk.)
52 weeks (wk.)   1 year (yr.)

(≈ means approximately equal to)

Metric System Of Weights And Measures

Units of Length

10 millimeters (mm) = 1 centimeter (cm)
10 centimeters (cm) = 1 decimeter (dm)
10 decimeters (dm) = 1 meter (m)
1,000 meters (m) = 1 kilometer (km)

Units of Capacity

1,000 milliliters (ml) = 1 liter (l)
1,000 liters (l) = 1 kiloliter (kl)

Units of Weight

1,000 milligrams (mg) = 1 gram (gm)
1,000 grams (gm) = 1 kilogram (kg)

Metric and United States Equivalents (approximate)

1 inch = 2.54 centimeters
1 meter = 39.37 inches
1 kilometer = .62 mile
1 liter = 1.1 quarts
1 gram = .034 ounces
1 kilogram = 2.2 pounds

1 we can work together

Everyone has the need to count and to use numbers. We get change in a store and count the money to be sure we have not been cheated. We read the thermometer to see how cold or warm it is. Our pay envelopes or checks, prices, amounts of food or materials, and many other examples are part of our everyday life.

Counting requires a system of making marks or using numbers. This chapter will show that a number system is also the basis for all the mathematics that follows. We refer to some old number systems but explain mostly the number system that is commonly used today.

A Plan To Save Time

There is a plan you can use to save time and get the most value from the chapter. You have just read an introduction which tells the general idea that will be covered. Next, see how much you already know and remember, by trying the pretest which follows. Check your answers with those at the end of the chapter, then review the sections which gave you trouble, or study the entire chapter carefully. Let's try it.

See What You Know And Remember — Pretest 1

Do as many of these questions or problems as you can. Some may be harder than others for you. Write your answers in the spaces provided.

1. All the different number symbols used in our number system are: ____________

2. Our number system is called a *decimal* system because it is based on: ____________

3. In a system of marking or tallying, how would you show *six* of something? ____________

4. If a large stone equals *ten* of something and a small stone equals *one* of something, what is the total number represented here? ____________

5. How many *digits* are in each of these numerals?

   a. 6 b. 29 ____ ____

   c. 8 d. 341 ____ ____

6. What *place* is shown by the underlined number

   a. in the numeral 63? b. in the numeral 59? ____ ____

7. In the following numerals, what *place* is shown by the underlined number?

   a. 427 b. 761,598 ____ ____

   c. 1,835 d. 96,240 ____ ____

   e. 6,952,064 ____

8. Write the numerals indicated by each of the following:

   a. 4 hundreds, 2 tens, 3 ones. ____________

   b. 2 thousands, 0 hundreds, 3 tens, 5 ones. ____________

   c. 5 thousands, 9 tens, 1 one. ____________

   d. 1 ten-thousand, 8 thousands, 3 hundreds, 6 tens, 7 ones. ____________

   e. 1 million, 0 hundred-thousands, 0 ten-thousands, 2 thousands, 5 hundreds, 9 tens, 0 ones. ____________

9. The numbers below follow a certain pattern. Fill in the two blanks on each line to continue the pattern.

a. 1	10	100	1,000	____	____
b. 30	300	3,000	30,000	____	____
c. 600,000	60,000	6,000	600	____	____
d. 1,500,000	150,000	15,000	1,500	____	____

10. In the Roman numeral system the symbol M represents 1,000, the symbol D represents 500, C is 100, L is 50, X is 10, V is five, and I is one. Write the following in Roman numerals.

a. 7 b. 28 ________ ________

c. 61 d. 620 ________ ________

e. 1,163 ________ ________

Write each of the following Roman numerals in our number system.

f. II g. XVIII ________ ________

h. CCLVI i. MDCCLXXVI ________ ________

Now turn to the end of the chapter to check your answers. Add up all that you had correct. Count by the number of separate answers, not by the number of questions. In this pretest there were 10 questions, but 37 separate answers.

A Score of	*Means That You*
34–37	Did very well. You can move to Chapter 2.
29–33	Know this material except for a few points. Read the sections about the ones you missed.
24–28	Need to check carefully on the sections you missed.
0–23	Need to work with this chapter to refresh your memory and improve your skills.

Questions	*Are Covered in Sections*
1, 2	1.1
3, 4	1.2
5	1.3
6–9	1.4 and 1.5
10	1.6

1.1 Number Symbols We Use

A *symbol* is a sign or device that stands for something else. We all know that a red traffic light means *stop*. The American flag represents the original thirteen states and the fifty states of today. A red cross ✚ means a hospital or medical help. Many other symbols are used to advertise products, tell the kinds of stores, show places on road maps, illustrate ideas in cartoons or pictures, and so forth. Two hands clasped together show cooperation. Two hands of the same person gripped overhead show victory. A fist means power or protest.

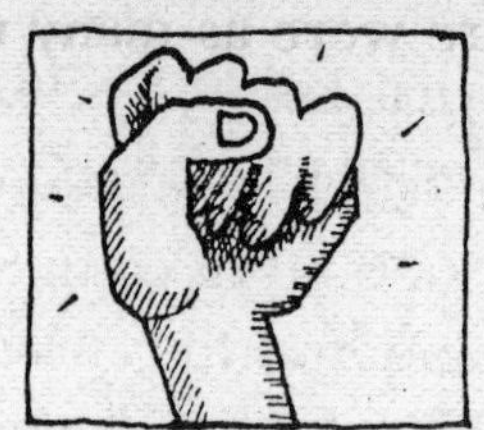

Number symbols represent an amount or quantity. For example, consider the number of toes an average person has on one foot. You could answer five, or if you spoke Spanish you could say *cinco*. Or you could write it as 5. The number or the amount of toes could also be shown as ~~IIII~~. Thus, the idea or *concept* of 5 can be shown in different ways. This will not apply to toes, but the *concept* of 5 can be shown as $3 + 2$, 1×5, $6 - 1$, $20 \div 4$, or by many other *numerals*. A *numeral* is a symbol used to denote a number. Thus, 5 is a numeral representing the number five.

In the same manner we can show any other amount or number by using only ten symbols: 0, 1, 2, 3, 4, 5, 6, 7, 8, 9. These symbols can be arranged to show any number that is needed. We must thank the Hindus for developing the symbols. The Arabs get credit for bringing the ideas to the Europeans, who brought them to our country. As a result, this number system is known as the Hindu-Arabic system. Because there are *ten* symbols and all numbers that we use are based on *ten*, we also call this a *decimal* system (*decimal* comes from the Roman word *decem*, which means *ten*). Later in this chapter we shall show more about the Roman and some other number systems.

1.2 Counting And Tallying

Many, many years ago, man kept a record of his flock of sheep by the use of number symbols. For each sheep he sent into the meadow, he made a notch in a stick. This notch was called a tally. There would be as many tallies or notches in his stick as there were sheep feeding in the meadow.

Man also used pebbles to count with. He kept a count of his sheep by placing one pebble in a pile for each sheep. The number of pebbles in the pile represented the number of sheep. This pile of pebbles represented six sheep:

For each method used, there were as many notches in a stick or pebbles in a pile as the number of things counted.

Each *one* of the pebbles or notches matches or corresponds with *one* of the objects. Therefore, we call this method of counting a *one-to-one correspondence.*

Years later, man devised a method to represent several objects by using a single symbol. He might have used a rock to replace each pile of five pebbles. We do the same today when we record data by tallying. We mark each *one* by a single stroke like I. We keep recording by single strokes until we have four strokes, IIII. The fifth mark is written as ~~IIII~~, thereby renaming that picture, five.

Imagine that you achieved a score of 29 on Pretest 1. This would be written as ~~IIII~~ ~~IIII~~ ~~IIII~~ ~~IIII~~ ~~IIII~~ IIII. How would you write your score using tally marks? Write it here. ______________________________

Written numerals developed when man traveled and traded his goods. He had a need for a more permanent record.

1.3 One-Digit, Two-Digit, And Larger Numbers

You know that the number symbols can be used to represent any number. When only *one* of the symbols is used, as in these examples,

a. 3 b. 9 c. 6 d. 4

we have a *single-digit* or a *one-digit* number. A *digit* is a single-number symbol.

A *two-digit* number would be similar to any of the following examples:

a. 21 b. 68 c. 59 d. 47

You see that two number symbols are put together to represent a larger amount. The same idea follows with *three-digit* numbers such as 613 or *four-digit* numbers such as 5,782 and so on as high as you want to go. However, the digits have different meaning according to their *place,* as you will learn in the next section.

It is interesting to note that *digit* comes from a Roman word which means *finger.* This shows a close connection between old (and still-used) methods of counting and the terms we use today.

Number Terms You Should Remember

Symbol A sign representing something else.

Decimal System A system of counting by 10.

Tally An account kept by notches.

Digit Any one of the Arabic numerals.

Numeral A symbol used to denote a number.

1.4 Each Place Has A Value

When you look at the numeral 2, you know it means *two* of something. How about the 2 in the numeral 23? It also means two of something. However, there is a big difference in the *somethings* that are represented.

Each digit takes on a different value, depending on the place it occupies. Thus the digit 2 stands for 2 tens and the 3 stands for 3 ones. This is shown in the following diagram.

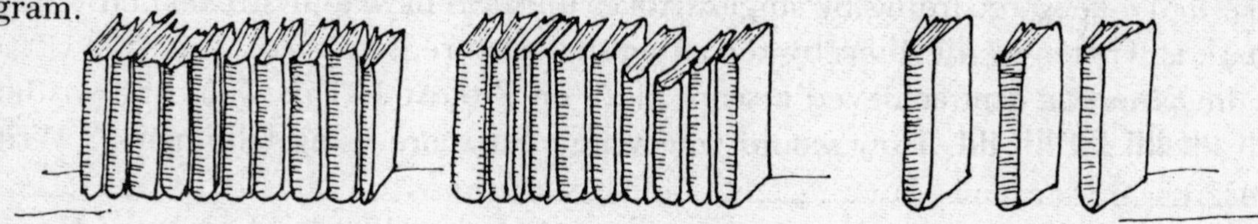

The number 23 has 2 tens and 3 ones =

TENS	ONES
2	3

You see that each digit in a number has a different value depending on its *place* in the number.

If you had a three-digit number like 537, the 5 occupies a new place called the hundreds place. Thus:

537 has 5 hundreds, 3 tens, and 7 ones =

HUNDREDS	TENS	ONES
5	3	7

See how place value permits you to write a number as large as you please without adding new symbols to the ten we have already.

Practice Exercise 1

Fill in the blanks for each of the following numbers.

1. 37 has ____________ tens and ____________ ones.
2. 73 has ____________ tens and ____________ ones.
3. 239 has ____________ hundreds and ____________ tens and ____________ ones.
4. 932 has ____________ hundreds and ____________ tens and ____________ ones.
5. 444 has ____________ hundreds and ____________ tens and ____________ ones.

Look at this numeral: 6,937

The digit 6 is in the fourth place to the left. This place tells how many *thousands* there are. Using a comma between the hundreds and thousands place makes the number easier to read. Commas are used after each group of three digits starting from the right. The comma in 6937 is placed between the 6 and the 9: 6,937. Thus, 6,937 has 6 thousands, 9 hundreds, 3 tens, and 7 ones.

6,937 =

THOUSANDS	HUNDREDS	TENS	ONES
6	9	3	7

If you regrouped the same four digits in a different way like 7,693, you would have a different numeral and hence an entirely different number. This is illustrated in the following diagram:

6,937 has 6 thousands, 9 hundreds, 3 tens, and 7 ones.

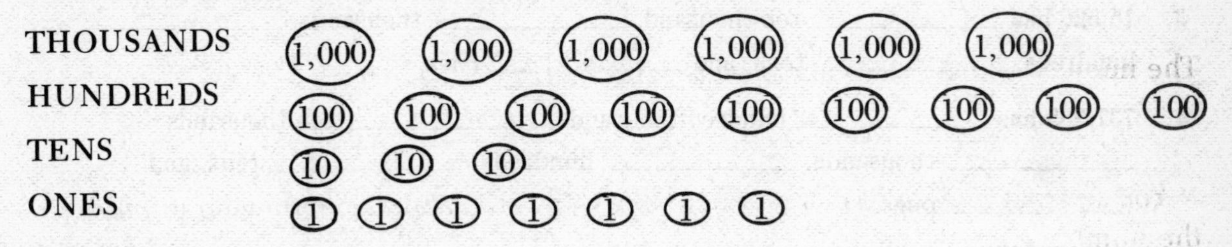

7,693 has 7 thousands, 6 hundreds, 9 tens, and 3 ones.

THOUSANDS (1,000) (1,000) (1,000) (1,000) (1,000) (1,000) (1,000)
HUNDREDS (100) (100) (100) (100) (100) (100)
TENS (10) (10) (10) (10) (10) (10) (10) (10) (10)
ONES (1) (1) (1)

The names for some of the other places are shown in the diagram below, using the number 3,237,471:

MILLIONS	HUNDRED-THOUSANDS	TEN-THOUSANDS	THOUSANDS
3	2	3	7

HUNDREDS	TENS	ONES
4	7	1

The number 3,237,471 has 3 millions, 2 hundred-thousands, 3 ten-thousands, 7 thousands, 4 hundreds, 7 tens, and 1 one. It is read as three million, two hundred thirty-seven thousand, four hundred seventy-one. Notice that the second comma is placed after the second group of three digits, or between the millions and the hundred-thousands.

Practice Exercise 2

Work this exercise carefully, and see if you can get all the answers correct.

1. In the numeral 578

a. what does the 8 mean? ________________

b. what does the 7 mean? ________________

c. what does the 5 mean? ________________

Fill in the missing blanks for each of these numbers:

2. 3,721 has __________ thousands, __________ hundreds, __________ tens, and __________ one.

3. 15,526 has __________ ten-thousand, __________ thousands, __________ hundreds, __________ tens, and __________ ones.

4. 737,985 has __________ hundred-thousands, __________ ten-thousands, __________ thousands, __________ hundreds, __________ tens, and __________ ones.

Write each of the following in numbers.

5. forty-six ________________

6. three hundred ninety-eight ________________

7. four thousand, two hundred nineteen ________________

8. twenty-one thousand, four hundred thirty-seven ________________

9. six hundred twenty-three thousand, one hundred twenty-one ________________

Write the numeral that means:

10. 5 tens and 3 ones. ________________

11. 3 hundreds, 4 tens, and 6 ones. ________________

Write each of the following numbers in written words.

12. 3,693 ________________________________

13. 726 ________________________________

14. 459,822 ________________________________

15. 8,333,333 ________________________________

16. Write the numeral, using all the digits 8, 6, 9, and 5, which represents:

a. the largest number. ________________

b. the smallest number. ________________

17. The four numbers below follow a pattern. Write the next two numbers that follow the same pattern.

a. 1   10   100   1,000   __________ __________

b. 30 300 3,000 30,000 ________ ________

c. 100,000 10,000 1,000 100 ________ ________

d. 600,000 60,000 6,000 600 ________ ________

1.5 Even Nothing Means Something

Zero is used as a *place holder*. If you want to write the number six hundred, you place the digit 6 in the hundreds place (the third place to the left) and follow it with zeros in the tens and ones places. Thus 600 is the numeral for the number six hundred. Without the use of the zeros, there would be no way to show that the digit 6 was in the hundreds place.

The symbol *zero* allows you to write a number as large as you wish. Years ago, before the invention of zero as a place holder man wrote 66 meaning many different numbers: sixty-six, six hundred sixty, six hundred six, and so on. With the invention of zero as a place holder, guessing the actual number is no longer necessary.

66 has 6 tens and 6 ones
606 has 6 hundreds, 0 tens, and 6 ones
660 has 6 hundreds, 6 tens, and 0 ones

Practice Exercise 3

1. Write the numeral that means

a. 4 tens and 0 ones. ________

b. 9 hundreds, 8 tens, and 0 ones. ________

c. 6 hundreds, 0 tens, and 1 one. ________

d. 7 thousands, 3 hundreds, and 4 ones. ________

e. 5 thousands and 6 tens. ________

2. Write as numerals

   a. Two thousand, thirty-two. ________

   b. Three thousand, sixty. ________

   c. Six hundred ninety. ________

   d. Six hundred ninety. ________

   e. Ten thousand. ________

1.6 Old Number Systems Still Influence Us

The movie *Chinatown* was released in MCMLXXIV. Do you know the year that represents?

Those numerals are Roman numerals, which formed the Roman number system. You still see these numbers used to name the hours on the faces of some clocks, to record dates on the cornerstone of public buildings, or to represent the year a movie (such as *Chinatown*) has been released. These are but a few places you see these numerals. Perhaps you could name many more places.

Each letter in the *Roman system* names a number. The basic Roman symbols and their values are listed in the following table.

Our Numeral	Roman Numeral Symbols
1	I
5	V
10	X
50	L
100	C
500	D
1,000	M

In this system, new numbers are formed by both addition and subtraction. For example,

$$\text{VIIII} = 5 + 4 = 9 \qquad \text{or} \qquad \text{IX} = 10 - 1 = 9$$

The four strokes after the five (*V*) means that four is to be *added* to the five; the stroke before the ten (*X*) means that 1 is to be *subtracted* from the ten. In each case, the result is 9.

Each form is acceptable but we usually look for the one which has fewer symbols. In this case IX is preferable.

Here are a few examples of Roman numerals illustrating the use of addition and subtraction to form new numbers.

Roman Numeral	Explanation	Hindu-Arabic* Numeral
XL	50 − 10. The X precedes the L so subtract ten from fifty.	40
LXXII	50 + 2 tens + 2 ones. L is fifty, XX is twenty, and II = 2 ones.	72
CCLXVI	200 + 50 + 10 + 5 + 1. CC = 200, L = 50, X = 10, V = 5, and I = 1.	266
DCXLI	500 + 100 + 50 − 10 + 1. D = 500, C = 100, XL = 50 − 10 = 40 and I = 1.	641
MDCXLIV	1,000 + 500 + 100 + 50 − 10 + 5 − 1. M = 1,000, D = 500, C = 100, XL = 50 − 10 = 40, IV = 5 − 1 = 4.	1,644

*System we use

In all number systems you are able to write a number of any size. Why is it to your advantage to use the Hindu-Arabic system rather than the older methods described?

Writing the same number each way will illustrate the *simplicity* of our number system. Take the number, one thousand five hundred forty-three. In Roman numerals it is written MDXLIII; and in our present system it is written 1,543.

As you see, fewer symbols are used to write a number in our Hindu-Arabic system in most cases than in the other system. This is especially true when writing large numbers. A simple number like 8, is written as VIII in the Roman system, can be expressed as a single symbol in our decimal system.

Other systems can be formed using a different number of digits. The Babylonians had a *base sixty system* which probably helped to create 60 minutes in an hour, 60 seconds in a minute, 360° (° means degrees) in a circle, and other measurements.

Practice Exercise 4

Show how well you understand Roman numerals by solving all these problems correctly. Use the chart on page 10 to figure out those you do not know.

1. What Roman numerals represent these numbers?

a. 67 ____________ f. 873 ____________

b. 105 ____________ g. 1,974 ____________

c. 345 ____________ h. 1,031 ____________

d. 508 ____________ i. 1,541 ____________

e. 101 ____________ j. 2,865 ____________

2. The following numerals are written in Roman numerals. Express these numbers in our decimal system.

a. LXII ____________

b. XLIV ____________

c. CXLIX ____________

d. CCLXVII ____________

e. DCXLI ____________

f. MCL ____________

g. CMLXI ____________

h. CDX ____________

i. MCMLXXIII ____________

j. MMCMIX ____________

3. The date on the cornerstone of a building reads MCMXII. What year does this represent? ____________

Terms You Should Remember

Numeral Symbol used to denote a number.

Place Value The value of the symbol 1, according to its place in the numeral (in 213, 1 has the value of ten).

Base The foundation on which the number system rests (decimal system — base 10).

Review Of Important Ideas

Some of the most important ideas in Chapter 1 were:

Numbers play a very important role in our life.

A numeral is a name for a number.

Each symbol in the numeral is the name of a number.

The position of the symbol in the numeral determines the value of the symbol.

Example: 742 = 7 hundreds, 4 tens, and 2 ones

The symbol *zero* represented by 0, holds a place that would otherwise be unoccupied.

Example: 502 = 5 hundreds, 0 tens, and 2 ones

The number of digits in a numeral determines the relative size of the number.

Example: A three-place numeral is larger than a two-place numeral.

Check What You Have Learned

The following test lets you see how well you have learned the ideas in Chapter 1. We call it a *post*test because it comes *after* the chapter. You noted that we called the test at the beginning a *pre*test because it came *before* the chapter. You can call them pretest and posttest, before and after tests, or whatever you like. The main point is to take the test to show yourself that you have learned the material.

Posttest 1

1. All the different number symbols used in our number system are: ______________

2. Our number system is called a decimal system because it is based on the number ___? ______________

3. In a system of tallying, how would you indicate five of something? ______________

4. If 𝍸 represents five of something and | represents one of something, what is the total number represented here? 𝍸 𝍸 𝍸 𝍸 || ______________

5. How many digits are in each of these numerals?

a. 38 ______________

b. 7 ______________

c. 101 ______________

d. 27 ______________

6. What place is shown by the underlined number?

a. 3$\underline{8}$ ______________

b. $\underline{3}$822 ______________

c. $\underline{8}$3213 ______________

d. $\underline{8}$ ______________

e. $\underline{9}$91 ______________

7. Fill in the blanks for each of the following numbers.

a. 38 = ______________ tens and ______________ ones.

b. 3,822 = ____________ thousands, ____________ hundreds, ____________ tens, and ____________ ones.

c. 83,213 = ____________ ten-thousands, ____________ thousands, ____________ hundreds, ____________ ten, and ____________ ones.

d. 8 = ____________ ones.

e. 991 = ____________ hundreds, ____________ tens, and ____________ one.

8. Write the numerals indicated by each of the following.

a. 5 hundreds, 3 tens, and 2 ones. ____________

b. 3 thousands, 0 hundreds, 3 tens, and 5 ones. ____________

c. 5 thousands, 9 hundreds, and 1 one. ____________

d. 1 ten-thousand, 8 thousands, 3 hundreds, 2 tens, and 5 ones. ____________

e. 2 millions, 0 hundred-thousands, 0 ten-thousands, 2 thousands, 5 hundreds, and 9 ones. ____________

9. In the Roman Numeral system, the symbol M represents 1,000, D = 500, C = 100, L = 50, X = 10, V = 5, and I = 1.

Write the following in Roman numerals.

a. 16 ____________

b. 97 ____________

c. 495 ____________

d. 1,974 ____________

Write each of the following Roman numerals in our decimal system.

e. XVIII ____________

f. XL ____________

g. MDCLXVII ____________

h. MCMLXXIV ____________

ANSWERS AND EXPLANATIONS TO POSTTEST 1

1. 0, 1, 2, 3, 4, 5, 6, 7, 8, 9

2. ten or 10

3. IIII

4. 22

5. a. 2 b. 1 c. 3 d. 2

6. a. tens b. thousands c. ten-thousands d. ones e. hundreds

7. a. 38 = *3* tens and *8* ones b. 3,822 = *3* thousands, *8* hundreds, *2* tens, and *2* ones

c. 83,213 = *8* ten-thousands, *3* thousands, *2* hundreds, *1* ten, and *3* ones d. 8 = *8* ones

e. 991 = *9* hundreds, *9* tens, and *1* one

8. a. 532
   b. 3,035
   c. 5,901
   d. 18,325
   e. 2,002,509

9. a. 16 = XVI
   b. 97 = XCVII
   c. 495 = CDXCV
   d. 1,974 = MCMLXXIV
   e. 10 + 5 + 3 = 18
   f. 50 − 10 = 40
   g. 1,000 + 500 + 100 + 50 + 10 + 5 + 2 = 1,667
   h. 1,000 + 1,000 − 100 + 50 + 20 + 5 − 1 = 1,974

In counting up your answers, remember that there were 31 separate answers in this test.

A Score of	*Means That You*
27–31	Did very well. You can move to Chapter 2.
24–26	Know this material except for a few points. Reread the sections about the ones you missed.
19–23	Need to check carefully on the sections you missed.
0–18	Need to review the chapter again to refresh your memory and improve your skills.

Questions	*Are Covered in Sections*
1, 2	1.1
3, 4	1.2
5	1.3
6–8	1.4 and 1.5
9	1.6

You Are On Your Way

You have made a good start. With a sound understanding of our number system you can learn to work with numbers. Chapter 2 will discuss addition, one of the basic operations with numbers.

Each individual must work out his or her own way to study and learn. Along the way we shall try to give ideas and methods that could help. Before you start Chapter 2, think about these points:

You study because you want to learn, not because somebody tells you to study.

We try to *suggest*, rather than tell. That leaves the choice up to you, as it should be.

A study suggestion — start with a goal. Have a reason for doing the study that is important to you. The goal or reason will pull you through the times when you get discouraged.

Mathematical Maze

This maze contains *seven* mathematical words. They are written horizontally →, vertically ↕, or diagonally ↗ ↘. When you locate a word in the maze, draw a ring around it. The word DIGIT has already been done for you.

S A N B C X D
R Y U Z Y H E
P O M S R L C
L J E B T T I
A K R A O A M
C I A S F L A
E T L E E L L
(D I G I T) Y O

WORDS OF:

4 letters
BASE

5 letters
PLACE
DIGIT
TALLY

6 letters
SYMBOL

7 letters
NUMERAL
DECIMAL

The solution is on page 18.

ANSWERS FOR CHAPTER 1

PRETEST 1

1. 0, 1, 2, 3, 4, 5, 6, 7, 8, 9

2. ten or 10

3. IIII I

4. 32

5. a. 1 b. 2 c. 1 d. 3

6. a. ones b. tens

7. a. hundreds b. hundred-thousands c. thousands d. ten-thousands e. millions

8. a. 423 b. 2,035 c. 5,091 d. 18,367 e. 1,002,590

9. a. 10,000 100,000 b. 300,000 3,000,000 c. 60 6 d. 150 15

10. a. VII
b. XXVIII
c. LXI
d. DCXX
e. MCLXIII
f. 2
g. 18
h. 256
i. 1,776

PRACTICE EXERCISE 1

1. 37 = 3 tens and 7 ones
2. 73 = 7 tens and 3 ones
3. 239 = 2 hundreds, 3 tens, and 9 ones
4. 932 = 9 hundreds, 3 tens, and 2 ones
5. 444 = 4 hundreds, 4 tens, and 4 ones

PRACTICE EXERCISE 2

1. a. 8 ones b. 7 tens c. 5 hundreds
2. 3,721 = 3 thousands, 7 hundreds, 2 tens and 1 one
3. 15,526 = 1 ten-thousand, 5 thousands, 5 hundreds, 2 tens, and 6 ones
4. 737,985 = 7 hundred-thousands, 3 ten-thousands, 7 thousands, 9 hundreds, 8 tens, and 5 ones
5. 46
6. 398
7. 4,219
8. 21,437
9. 623,121
10. 53
11. 346
12. three thousand, six hundred ninety-three
13. seven hundred twenty-six
14. four hundred fifty-nine thousand, eight hundred twenty-two
15. eight million, three hundred thirty-three thousand, three hundred thirty-three
16. a. 9,865 b. 5,689
17. a. 10,000 100,000 b. 300,000 3,000,000 c. 10 1 d. 60 6

PRACTICE EXERCISE 3

1. a. 40 b. 980 c. 601 d. 7,304 e. 5,060
2. a. 2,032 b. 3,060 c. 690 d. 40,310 e. 10,000

PRACTICE EXERCISE 4

1. a. LXVII
   b. CV
   c. CCCXLV
   d. DVIII
   e. CI
   f. DCCCLXXIII
   g. MCMLXXIV
   h. MXXXI
   i. MDXLI
   j. MMDCCCLXV

2. a. $50 + 10 + 2 = 62$
b. $50 - 10 + 5 - 1 = 44$
c. $100 + 50 - 10 + 10 - 1 = 149$
d. $200 + 50 + 10 + 5 + 2 = 267$
e. $500 + 100 + 50 - 10 + 1 = 641$
f. $1{,}000 + 100 + 50 = 1{,}150$
g. $1{,}000 - 100 + 50 + 10 + 1 = 961$
h. $500 - 100 + 10 = 410$
i. $1{,}000 + 1{,}000 - 100 + 50 + 20 + 3 = 1{,}973$
j. $2{,}000 + 1{,}000 - 100 + 10 - 1 = 2{,}909$

3. MCMXII $= 1{,}000 + 1{,}000 - 100 + 10 + 2 = 1912$

MATHEMATICAL MAZE SOLUTION

In the last chapter you learned about number systems. Chapter 2 starts to talk about things you can *do* with numbers. In other words, this discusses how to *use* numbers or how to *operate* with them.

The topic is *addition,* which is a basic operation or use of numbers. You know you can figure the total of two or more numbers by adding them together. No matter how large or how long an addition example may seem, it can be done by following certain basic steps. Check your skill on this topic in the pretest which follows.

See What You Know And Remember — Pretest 2A

Work these exercises carefully, doing as many problems as you can. Write the answers below each question.

1. $\begin{array}{r} 4 \\ +3 \\ \hline \end{array}$ **2.** $\begin{array}{r} 6 \\ +0 \\ \hline \end{array}$ **3.** $\begin{array}{r} 34 \\ +\ 4 \\ \hline \end{array}$ **4.** $\begin{array}{r} 42 \\ +25 \\ \hline \end{array}$ **5.** $\begin{array}{r} 2 \\ 3 \\ +4 \\ \hline \end{array}$

6.
$$\begin{array}{r} 242 \\ +325 \\ \hline \end{array}$$

7.
$$\begin{array}{r} 257 \\ +\ 32 \\ \hline \end{array}$$

8. $7 + 2 =$

9.
$$\begin{array}{r} 123 \\ 12 \\ 210 \\ +\ 24 \\ \hline \end{array}$$

10.
$$\begin{array}{r} 31 \\ 12 \\ +23 \\ \hline \end{array}$$

11. Find the sum of 341, 2, 20, 401, and 133.

12. How many clams would I have if I dug up 12 on Monday, 21 on Tuesday, 10 on Wednesday, and 14 on Thursday?

13. There were 221 qts. of milk sold during the first hour, 233 qts. during the second hour, 320 qts. during the third hour, and only 121 during the fourth hour. What was the total number of quarts sold during the four-hour period?

14. There are 4,032 men working for the Transit Authority on the night shift and 25,146 men on the day shift. What is the total number of men employed?

15. In the parentheses in each example, insert one of these symbols to make a true statement:

$>$ which means *more than*
$=$ which means *the same*
$<$ which means *less than*

a. 4( )7 b. 6( )4 c. 7( )7

d. 15( )13 e. 1( )3

Now turn to page 36 to check your answers. Add up all that you had correct. Count by the number of separate answers, not by the number of questions. In this pretest there were 15 questions but 19 separate answers.

A Score of	*Means That You*
18–19	Did very well. You can move to the second part of this chapter, "The Sum of Things."
15–17	Know this material except for a few points. Read the sections about the ones you missed.
12–14	Need to check carefully on the sections you missed.
0–11	Need to work with this part of the chapter to refresh your memory and improve your skills.

Questions	*Are Covered in Sections*
1, 4, 6, 8	2.2
15	2.3
2	2.4
3, 7	2.5
5, 9–11	2.6
12–14	2.7

2.1 Whole Numbers

The numbers we generally use are called *whole numbers*. This term is used because there are other examples which we will learn later that are parts of a whole. They are called *fractions* and *decimals*. For now and the next few chapters we are working with whole numbers. These are the numbers:

$$0, 1, 2, 3, 4, 5, 6, 7, 8, 9$$

and all the higher numbers made by combining two or more of these symbols.

2.2 The Language Of Addition

Example

Mr. Lewis paid $4 to the baby-sitter on Thursday evening and $3 to the same sitter on Friday evening. How much money did he give to the baby-sitter for the two nights of work?

Solution

To find out what the baby-sitter earned for the two nights, add $4 and $3. Written in column form, the two *addends* look like this:

To show that you must add, use a *plus sign*.

$$\begin{array}{r} \$4 \\ +\$3 \\ \hline \$7 \end{array}$$

The *sum* or *total* of $4 and $3 is shown here.

You notice that addition has a language. Or you could say there are terms you must know to understand addition.

Terms You Should Remember

Add To join to another.

Addend One of the quantities that is to be added to another.

Sum The result obtained when two or more numbers are added.

Plus The sign (+) to denote addition.

Column A single vertical file.

Practice Exercise 5

Find the sum of each of the following problems. Show how well you can do by getting all 36 correct. Answers to this exercise and the following tests begin on page 36.

1. $\begin{array}{r}3\\+6\\\hline\end{array}$	**2.** $\begin{array}{r}1\\+1\\\hline\end{array}$	**3.** $\begin{array}{r}2\\+3\\\hline\end{array}$	**4.** $\begin{array}{r}7\\+2\\\hline\end{array}$	**5.** $\begin{array}{r}1\\+3\\\hline\end{array}$	**6.** $\begin{array}{r}5\\+1\\\hline\end{array}$
7. $\begin{array}{r}3\\+4\\\hline\end{array}$	**8.** $\begin{array}{r}5\\+2\\\hline\end{array}$	**9.** $\begin{array}{r}2\\+2\\\hline\end{array}$	**10.** $\begin{array}{r}1\\+5\\\hline\end{array}$	**11.** $\begin{array}{r}5\\+3\\\hline\end{array}$	**12.** $\begin{array}{r}1\\+7\\\hline\end{array}$
13. $\begin{array}{r}3\\+2\\\hline\end{array}$	**14.** $\begin{array}{r}4\\+2\\\hline\end{array}$	**15.** $\begin{array}{r}6\\+2\\\hline\end{array}$	**16.** $\begin{array}{r}3\\+3\\\hline\end{array}$	**17.** $\begin{array}{r}4\\+5\\\hline\end{array}$	**18.** $\begin{array}{r}8\\+1\\\hline\end{array}$
19. $\begin{array}{r}3\\+5\\\hline\end{array}$	**20.** $\begin{array}{r}4\\+1\\\hline\end{array}$	**21.** $\begin{array}{r}7\\+1\\\hline\end{array}$	**22.** $\begin{array}{r}5\\+4\\\hline\end{array}$	**23.** $\begin{array}{r}6\\+1\\\hline\end{array}$	**24.** $\begin{array}{r}2\\+1\\\hline\end{array}$

25.	26.	27.	28.	29.	30.
1 +8	2 +7	4 +4	1 +4	2 +4	6 +3

31.	32.	33.	34.	35.	36.
2 +5	1 +2	4 +3	2 +6	3 +1	1 +6

Example

Mrs. Patrick is employed as a domestic worker. She earned $24 on Monday and $22 on Tuesday. How much money did she earn altogether?

Solution

To find out what Mrs. Patrick earned you must add her earnings for the two days. Add $24 and $22 like this:

Problem	Step 1	Step 2
$24 + $22	$ 2 4 + $ 2 2 6	$ 2 4 + $ 2 2 $ 4 6
	Add the *ones* place first: 4 + 2 = 6	Add the *tens* place next: 2 + 2 = 4

The sum of $24 + $22 = $46.

Practice Exercise 6

Add each of the following.

1.	2.	3.	4.	5.
35 +23	12 +47	25 +24	53 +25	71 +11

6.	7.	8.	9.	10.
83 +15	33 +25	71 +28	65 +24	56 +31

Example

Union dues are collected each month from Mrs. Ryan's paycheck. If she paid $112 in January and $123 in February, what was the total money deducted from her salary for the two months?

Solution

Once again you must add $112 and $123 to find the total deducted.

Problem	Step 1	Step 2	Step 3
$112 +$123	$ 1 1 [2] + $ 1 2 [3] [5]	$ 1 [1] 2 + $ 1 [2] 3 [3] 5	$ [1] 1 2 + $ [1] 2 3 $ [2] 3 5
	Add the *ones* place: 2 + 3 = 5	Add the *tens* place: 1 + 2 = 3	Add the *hundreds* place: 1 + 1 = 2

The sum of $112 + $123 = $235.

Practice Exercise 7

Add.

1. 235 + 342
2. 346 + 621
3. 712 + 273
4. 823 + 166
5. 526 + 253
6. 511 + 345
7. 387 + 412
8. 452 + 311
9. 313 + 316
10. 621 + 271

This same method is applied when you add numbers of more than three places so try the next practice exercise to test your skill.

Practice Exercise 8

Add.

1. 3,437 + 2,521
2. 35,432 + 23,517
3. 123,456 + 345,212
4. 35,382 + 23,412
5. 5,383,112 + 3,112,583

Numbers are written horizontally sometimes to save space but it is easier to rewrite the problem in column form and add as you did before.

Problem	Step 1	Step 2	Step 3
33 + 24 =	33 +24	3 3 + 2 4 7	3 3 + 2 4 5 7
	Rewrite the problem in column form	Add the *ones* place: 3 + 4 = 7	Add the *tens* place: 3 + 2 = 5

Thus: 33 + 24 = 57.

Practice Exercise 9

Rewrite each of the following problems in column form first and then add.

1. 41 + 45 =

2. 42 + 32 =

3. 64 + 11 =

4. 73 + 23 =

5. 85 + 13 =

6. 21 + 31 =

7. 12 + 71 =

8. 66 + 23 =

9. 52 + 24 =

10. 41 + 53 =

2.3 Help From The Number Line

It is sometimes convenient to represent *numbers* by points on a line. For example, draw the line below, and assign the value zero to any point you wish. Then, along the line at any convenient distance from zero, represent the number one. Using the same distance from zero to one as your measuring unit, mark off the points associated with two, three, four . . . (the three dots [ellipsis] indicate that the line may go on indefinitely). There is no end to the line or numbers on the line. A line such as this is called a *number line*.

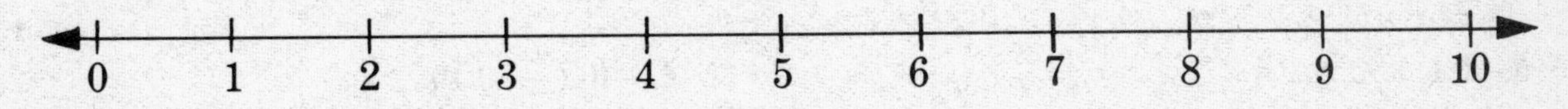

Using the number line, you can find the number of units between 3 and 7. Start at 3 and move to the right from A to B as shown below.

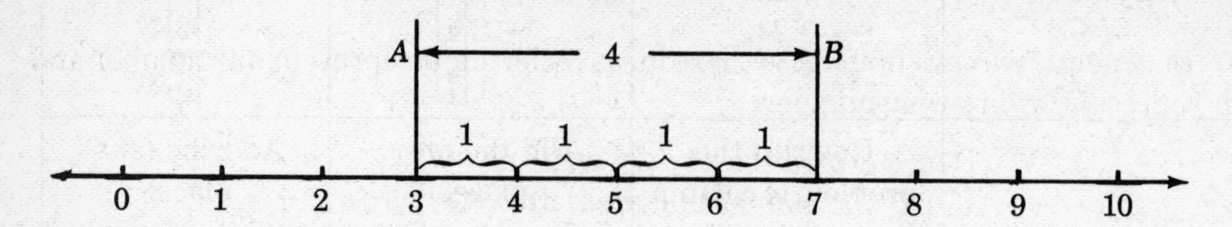

Thus you see that there are *four* units between 3 and 7.

You can also see that any number to the *right* of another has the greater numerical value. Thus 7, which is to the right of 3, is the greater number. This is indicated by a new symbol, $>$, which means *greater than.* Since 7 is *greater than* 3, this statement is written as $7 > 3$.

Similarly, 9 is to the right of 7 because 9 is *greater than* 7. This statement is written as $9 > 7$.

These are examples of *inequalities.* An *inequality* is a statement that two amounts are not equal.

Any number to the left of another has a smaller numerical value. Because 5 is to the left of 8, 5 is the smaller number. This is indicated by the symbol $<$, *less than.* Since 5 is *less than* 8, it is written as $5 < 8$.

The number 8 is to the left of 10, so 8 is *less than* 10. This is written: $8 < 10$. These are also examples of inequalities.

A number can be equal to, greater than, or less than another number. Once again, the symbols are:

greater than	$>$
equal to	$=$
less than	$<$

Practice Exercise 10

In the parentheses between the numbers insert the symbol $>$, $<$, or $=$ which will make a true statement. The first one has been done for you.

1. 7 ($>$) 6
7 is greater than 6

2. 5 (   ) 2

3. 3 (   ) 8

4. 6 (   ) 6

5. 1 (   ) 2

6. 10 (   ) 10

7. 5 ( ) 7 8. 9 ( ) 3

9. 8 ( ) 8 10. 4 ( ) 5

In general, we can summarize these ideas by letting a represent one number and b represent a different number.

If a is *larger* than b, we write $a > b$.
If a is the same as or *equal* to b, we write $a = b$.
If a is *less* than b, we write $a < b$.

Practice Exercise 11

Try this exercise to see how well you understand these ideas. Answers are found on page 37.

1. Using the horizontal line below, draw a number line labeling it from 0 to 10.

2. What number is 3 units to the right of 3? ______

3. What number is 1 unit to the right of 8? ______

4. Count 3 units to the right of 2. What number corresponds to this point? ______

5. Count 2 units to the left of 8. What number corresponds to this point? ______

6. What number is 4 units to the right of 8? (The number line may be extended, if necessary.) ______

7. Which is the larger number, 6 or 5? Write this, using the inequality symbol. ______

8. The temperature on Thursday morning was 45° and in the afternoon was 52°. How many degrees did it rise? (*Hint:* Picture the number line in your head when working with large numbers.) ______

9. The price of an automobile rose from $3,482 to $3,491. How much money has it risen? ______

Adding whole numbers can also be shown on the number line. Let's look again at Mr. Lewis's baby-sitting problem in Section 2.2. He paid $4 one night and $3 the next night. To find out how much was paid, we added the two amounts of money together. Adding $4 and $3 on the number line can be shown in the following way.

To begin, represent 4 by a line segment starting at 0 and extending 4 units to the right of 0. Remember 0 is our starting point.

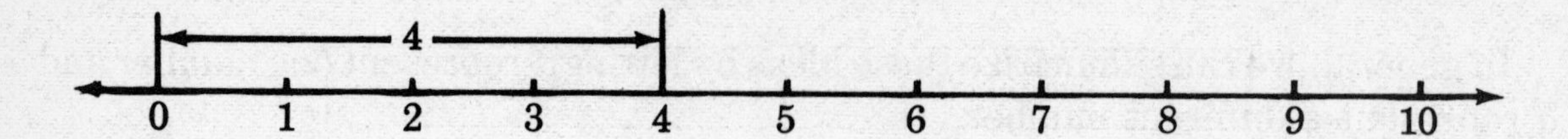

Represent 3 (the second addend) by a line segment starting at 4, the end of the first line segment, and extending 3 units to the right.

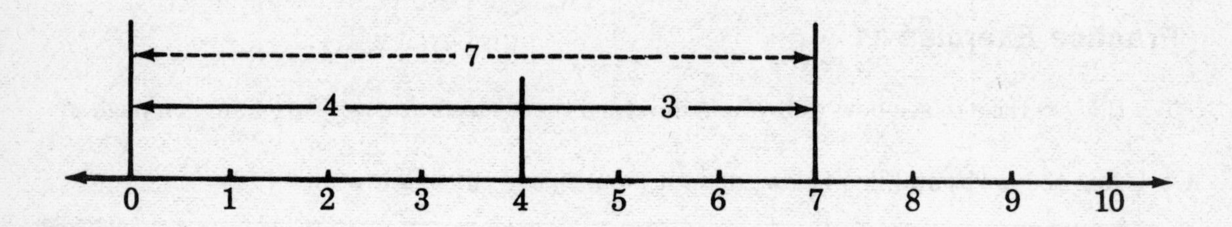

The second segment ends at a point marked 7 on the number line. This point can also be reached with a single line segment starting and 0 and extending 7 units to the right. Thus the dotted segment represents the sum of the other two segments. The sum of 4 and 3, written as 4 + 3, is equivalent to 7. Thus:

$4 + $3 = $7.

Practice Exercise 12

Using the number line given for each exercise, illustrate the addition indicated by each problem. First represent each portion; then use a dotted line to show the single line segment. Refer to the example just shown.

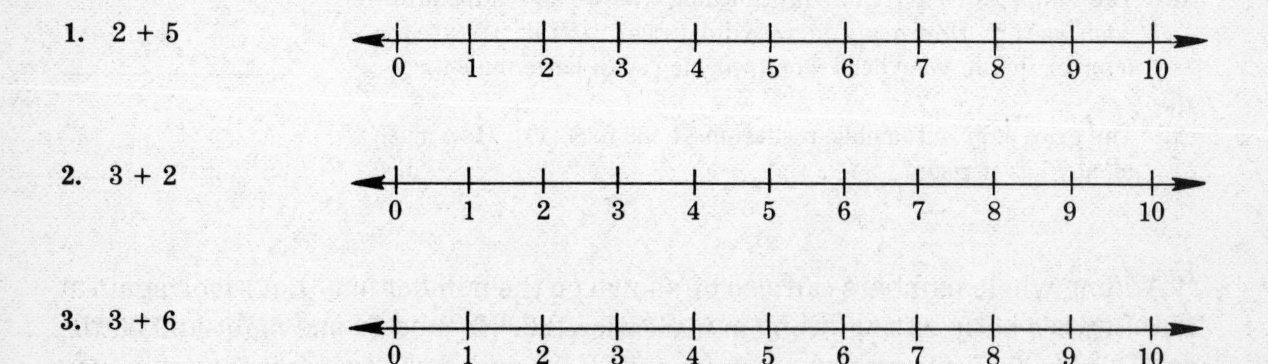

2.4 Nothing Is Still Important

Adding a one-place number and zero is done in this way:

Problem	Step 1
$\begin{array}{r} 3 \\ +0 \\ \hline \end{array}$	$\begin{array}{r} 3 \\ +0 \\ \hline \end{array}$
	Zero added to any number results in the number itself: $3 + 0 = 3$

Practice Exercise 13

Add.

1. $\begin{array}{r} 8 \\ +0 \\ \hline \end{array}$ 2. $\begin{array}{r} 4 \\ +0 \\ \hline \end{array}$ 3. $\begin{array}{r} 0 \\ +7 \\ \hline \end{array}$ 4. $\begin{array}{r} 5 \\ +0 \\ \hline \end{array}$ 5. $\begin{array}{r} 3 \\ +0 \\ \hline \end{array}$

6. $\begin{array}{r} 0 \\ +9 \\ \hline \end{array}$ 7. $\begin{array}{r} 0 \\ +3 \\ \hline \end{array}$ 8. $\begin{array}{r} 1 \\ +0 \\ \hline \end{array}$ 9. $\begin{array}{r} 6 \\ +0 \\ \hline \end{array}$ 10. $\begin{array}{r} 0 \\ +7 \\ \hline \end{array}$

11. $\begin{array}{r} 0 \\ +6 \\ \hline \end{array}$ 12. $\begin{array}{r} 0 \\ +8 \\ \hline \end{array}$ 13. $\begin{array}{r} 0 \\ +1 \\ \hline \end{array}$ 14. $\begin{array}{r} 0 \\ +2 \\ \hline \end{array}$ 15. $\begin{array}{r} 9 \\ +0 \\ \hline \end{array}$

16. $\begin{array}{r} 2 \\ +0 \\ \hline \end{array}$ 17. $\begin{array}{r} 0 \\ +5 \\ \hline \end{array}$ 18. $\begin{array}{r} 0 \\ +4 \\ \hline \end{array}$ 19. $\begin{array}{r} 0 \\ +0 \\ \hline \end{array}$

Example

A container of milk cost 75¢ before it increased 10¢ in price. What is its present cost?

Solution

You add 75¢ and 10¢:

Problem	Step 1	Step 2
75¢ +10¢	7 [5] ¢ + 1 [0] ¢ [5] ¢	[7] 5 ¢ + [1] 0 ¢ [8] 5 ¢
	Add the *ones* place: 5 + 0 = 5	Add the *tens* place: 7 + 1 = 8

Thus: 75¢ + 10¢ = 85¢.

Practice Exercise 14

Add these problems.

1. 53 +20
2. 408 +151
3. 360 +127
4. 532 +230
5. 646 +200
6. 3,003 +2,402
7. 30,000 +23,452
8. 3,553,222 +2,220,037
9. 6,903 +1,002
10. 403,246 +190,030

2.5 Unequal Place Addition

Example

At the end of the sixth frame in a bowling game, Jim's score is 142 pins. In the seventh frame he scores 6 pins more. How many pins does he now have to his credit?

Solution

Add 142 and 6 to find his score. This requires you to add a one-place number to a three-place number. To do this, consider the number in the ones place as having 0 hundreds and 0 tens. It would appear like this:

Problem	Step 1	Step 2	Step 3
142 + 6 ———	1 4 [2] + [6] ——— [8]	1 [4] 2 + [0] 6 ——— [4] 8	[1] 4 2 + [0] 0 6 ——— [1] 4 8
	Add: 2 + 6 = 8	Add: 4 + 0 = 4	Add: 1 + 0 = 1

Thus: 142 + 6 = 148.

Practice Exercise 15

Find these sums.

1. $\begin{array}{r} 35 \\ +\ 4 \\ \hline \end{array}$

2. $\begin{array}{r} 308 \\ +\ 41 \\ \hline \end{array}$

3. $\begin{array}{r} 546 \\ +\ 52 \\ \hline \end{array}$

4. $\begin{array}{r} 4{,}332 \\ +\ 146 \\ \hline \end{array}$

5. $\begin{array}{r} 6{,}721 \\ +\ 57 \\ \hline \end{array}$

6. $\begin{array}{r} 387 \\ +\ 12 \\ \hline \end{array}$

7. $\begin{array}{r} 43 \\ +\ 6 \\ \hline \end{array}$

Write in column form before adding.

8. 483 + 12 =

9. 5,432 + 27 =

10. 3,432,703 + 4,221 =

2.6 No Addition Problem Is Long

No matter how long a column of numbers may be, the addition process basically involves two numbers being added at one time. For example: you really have a series of small problems of addition.

$\begin{array}{r} 6 \\ 1 \\ +2 \\ \hline 9 \end{array}$ (6 and 1 → 7; 7 and 2 → 9)

Step 1: 6 + 1 = 7

Step 2: 7 + 2 = 9

Practice Exercise 16

Write each of these examples in column form and then add.

1. 3 + 2 + 4 =
2. 5 + 0 + 3 =
3. 2 + 2 + 3 + 4 =
4. 2 + 1 + 4 + 1 =
5. 6 + 0 + 1 + 0 + 2 =

Numbers with more than a ones place are added the same way. Look at this problem.

Problem	Step 1	Step 2
43 22 +14	43 22 + 14 9	43 22 + 14 79
	Add the *ones* place: 3 + 2 + 4 = 9	Add the *tens* place: 4 + 2 + 1 = 7

Thus: 43 + 22 + 14 = 79.

2.7 Word Problems

Word problems introduce new ideas in this book. Whenever possible, this type of question will be used to show you the wording involved in the particular operation.

General Instructions

1. Read the problem carefully. Reread it again to be sure you understand it.
2. Discover the information that is told you or what is *given*.
3. Find out what question you are being asked.
4. Determine the *method* you need to work the problem. Should you add, subtract, multiply, divide?
5. Finally, *check* to see if your answer fits. Did you answer the question asked?

Key Words Or Phrases For Problems Solved By Adding

1. *Sum* or *total* — "What is the *sum* or *total* money deducted from his salary for the two months?"
2. *Increased* — "An item *increased* 10¢ in price."

3. *More than* — "I have 5¢ *more than* I had yesterday."
4. *Added to* — "Sixteen pens were *added to* the 142 I had before."

Practice Exercise 17

Search for the key words and solve the problems.

1. On Tuesday 2 bottles of milk were bought, on Wednesday 3 bottles, and on Thursday 1 bottle. How many bottles were purchased altogether?

2. There were 43 cans of vegetables sold one day and 22 the next day in the L & Z store. What was the total number of cans sold during the two-day period?

3. Membership in the museum was increased $5 from the previous rate of $35. What is the new membership cost?

4. The New Mexico Giants added 3 more runs to the 4 they had already. How many runs do they have?

5. It is 4 blocks to the bowling alley, 3 more to the pizza parlor, and 2 more to the candy store. How many blocks is it to the candy store?

6. Last year there were 103 people on the day shift and only 46 on the night shift. How many people were employed altogether?

7. No more than 8 items purchased permits you to use the express lane in the supermarket. If I buy 3 bottles of soda, 1 bar of butter, 2 cans of tomato juice, 1 can of cleanser, and 2 packages of frozen vegetables, may I use the express lane?

8. Tickets for illegal parking are issued each day at the town hall parking area. What was the total number of tickets issued for the first three days in August if 41, 23, and 14 summonses were given?

9. Your earnings were $4,324 last year and your wife's earnings were $2,435. What were your combined earnings?

10. In a neighborhood street fair $5,035 was collected on Saturday and $1,703 more on Sunday. What was the total money collected by Sunday?

Terms You Should Remember

Equal Having the same value.

Unequal Not having the same value.

Number Line Points on a line, used to represent our number system.

Review Of Important Ideas

Some of the most important ideas in this part of Chapter 2 were:

The numbers we are adding are called addends.

The symbol + means to add.

The same place values are lined up in columns before adding.

Numbers are added by pairs only.

Let's Check How Well You're Doing

This section of the chapter moved from simple to more advanced addition. You probably found much of this coming back to you from the last time you used it. Posttest 2A lets you check on your understanding. Try these examples. The answers follow the posttest.

Posttest 2A

Write your answers below each problem.

1. $\begin{array}{r} 5 \\ +3 \\ \hline \end{array}$ **2.** $\begin{array}{r} 4 \\ +0 \\ \hline \end{array}$ **3.** $\begin{array}{r} 43 \\ +\ 3 \\ \hline \end{array}$ **4.** $\begin{array}{r} 55 \\ +23 \\ \hline \end{array}$ **5.** $\begin{array}{r} 4 \\ 1 \\ +3 \\ \hline \end{array}$

6. $\begin{array}{r} 523 \\ +245 \\ \hline \end{array}$ **7.** $\begin{array}{r} 457 \\ +\ 41 \\ \hline \end{array}$ **8.** $6 + 3 =$ **9.** $\begin{array}{r} 213 \\ 14 \\ 110 \\ +\ 52 \\ \hline \end{array}$ **10.** $\begin{array}{r} 41 \\ 14 \\ +22 \\ \hline \end{array}$

11. Find the sum of 143, 2, 30, 401, and 123.

12. Twelve oysters were found on the beach on Monday, 15 on Tuesday, 20 on Wednesday, and 21 on Thursday. What was the total of all oysters found?

13. There were 222 cups of coffee sold during the first hour, 233 cups in the second hour, 403 cups during the third hour, and only 120 cups during the fourth hour. What was the total number of cups of coffee sold during this four-hour period?

14. There were 43,622 elementary school teachers and 6,257 secondary school teachers in the year 1969. What was the total number of teachers employed that year?

15. In the parentheses, place the symbol >, =, or < to make a true statement.

a. 6 (   ) 4 b. 8 (   ) 3 c. 1 (   ) 10

d. 9 (   ) 9 e. 13 (   ) 15

ANSWERS AND EXPLANATIONS TO POSTTEST 2A

1. 8
2. 4
3. 46
4. 78
5. 8
6. 768
7. 498
8. 9
9. 389
10. 77
11. $\begin{array}{r} 143 \\ 2 \\ 30 \\ 401 \\ +123 \\ \hline 699 \end{array}$
12. $\begin{array}{r} 12 \\ 15 \\ 20 \\ +21 \\ \hline 68 \end{array}$
13. $\begin{array}{r} 222 \\ 233 \\ 403 \\ +120 \\ \hline 978 \end{array}$
14. $\begin{array}{r} 43{,}622 \\ +\ 6{,}257 \\ \hline 49{,}879 \end{array}$
15. a. 6 (>) 4 b. 8 (>) 3 c. 1 (>) 10
    d. 9 (=) 9 e. 13 (<) 15

In counting up your answers, remember that there were 19 separate answers in this test.

A Score of	*Means That You*
18–19	Did very well. You can move to the second part of this chapter, "The Sum of Things."
15–17	Know this material except for a few points. Reread the sections about the ones you missed.
12–14	Need to check carefully on the sections you missed.
0–11	Need to review this part of the chapter again to refresh your memory and improve your skills.

Questions	*Are Covered in Section*
1, 4, 6, 8	2.2
15	2.3
2	2.4
3, 7	2.5
5, 9–11	2.6
12–14	2.7

ANSWERS FOR CHAPTER 2—A

PRETEST 2A

1. 7   2. 6   3. 38   4. 67   5. 9

6. 567   7. 289   8. 9   9. 369   10. 66

11. 897   12. 57   13. 895   14. 29,178

15. a. $4 < 7$ b. $6 > 4$ c. $7 = 7$ d. $15 > 13$ e. $1 < 3$

PRACTICE EXERCISE 5

1. 9   2. 2   3. 5   4. 9   5. 4   6. 6

7. 7   8. 7   9. 4   10. 6   11. 8   12. 8

13. 5   14. 6   15. 8   16. 6   17. 9   18. 9

19. 8   20. 5   21. 8   22. 9   23. 7   24. 3

25. 9   26. 9   27. 8   28. 5   29. 6   30. 9

31. 7   32. 3   33. 7   34. 8   35. 4   36. 7

PRACTICE EXERCISE 6

1. 58   2. 59   3. 49   4. 78   5. 82

6. 98   7. 58   8. 99   9. 89   10. 87

PRACTICE EXERCISE 7

1. 577   2. 967   3. 985   4. 989   5. 779

6. 856   7. 799   8. 763   9. 629   10. 892

PRACTICE EXERCISE 8

1. 5,958   2. 58,949   3. 468,668   4. 58,798

5. 8,495,695

PRACTICE EXERCISE 9

1. 86   2. 74   3. 75   4. 96   5. 98

6. 52   7. 83   8. 89   9. 76   10. 94

PRACTICE EXERCISE 10

1. $7 > 6$ 2. $5 > 2$ 3. $3 < 8$ 4. $6 = 6$ 5. $1 < 2$

6. $10 = 10$ 7. $5 < 7$ 8. $9 > 3$ 9. $8 = 8$ 10. $4 < 5$

PRACTICE EXERCISE 11

1.

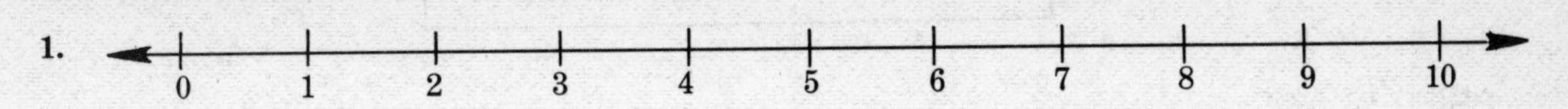

2. 6	3. 9	4. 5	5. 6
6. 12	7. 6, $6 > 5$	8. 7°	9. $9

PRACTICE EXERCISE 12

1.

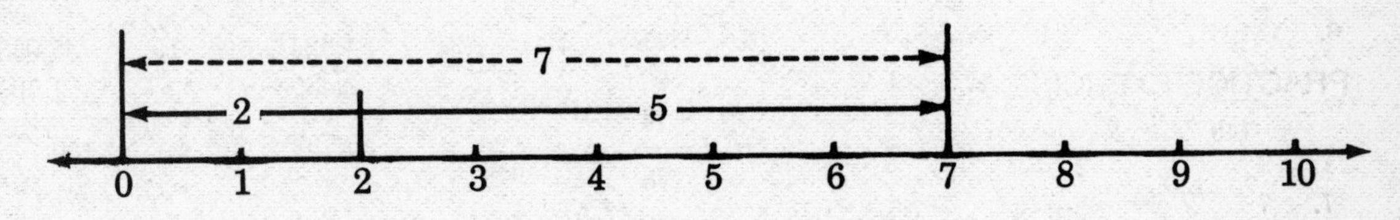

2.

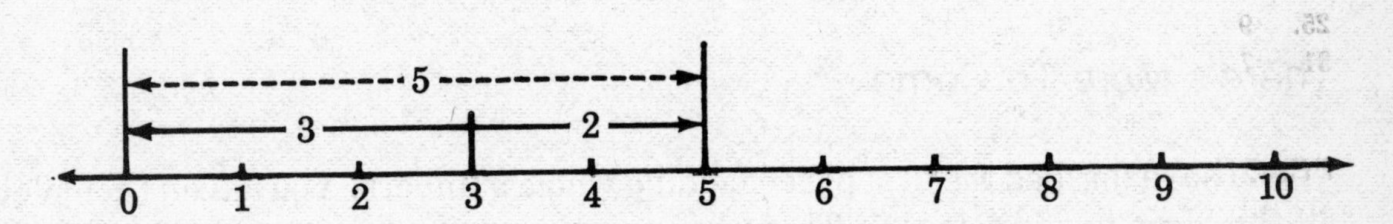

3.

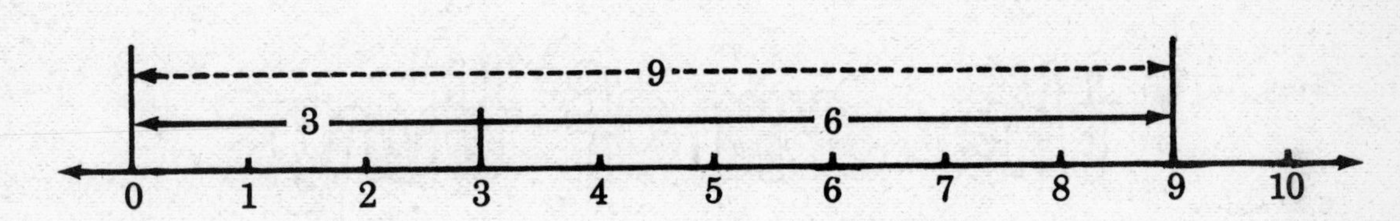

PRACTICE EXERCISE 13

1. 8	2. 4	3. 7	4. 5	5. 3
6. 9	7. 3	8. 1	9. 6	10. 7
11. 6	12. 8	13. 1	14. 2	15. 9
16. 2	17. 5	18. 4	19. 0	

PRACTICE EXERCISE 14

1. 73	2. 559	3. 487	4. 762	5. 846
6. 5,405	7. 53,452	8. 5,773,259	9. 7,905	10. 593,276

PRACTICE EXERCISE 15

1. 39	2. 349	3. 598	4. 4,478	5. 6,778
6. 399	7. 49	8. 495	9. 5,459	10. 3,436,924

PRACTICE EXERCISE 16

1. 9 **2.** 8 **3.** 11 **4.** 8 **5.** 9

PRACTICE EXERCISE 17

1. 2 3 +1 ——— 6	**2.** 43 +22 ——— 65	**3.** $35 + 5 ——— $40	**4.** 4 +3 ——— 7	**5.** 4 3 +2 ——— 9
6. 103 + 46 ——— 149	**7.** No. 3 1 2 1 +2 ——— 9	**8.** 41 23 +14 ——— 78	**9.** $4,324 + 2,435 ——— $6,759	**10.** $5,035 + 1,703 ——— $6,738

There's More To Come

You have finished the first part of adding whole numbers. You haven't completed all the ideas in addition yet. There is more to come.

the sum of things

Numbers come in many sizes. You have to learn to add many numbers together and you have to learn to add numbers that have large digits. Each type of example in this half of the chapter uses all that you have learned so far in Chapter 2, as well as some new ideas. Check your skill on these new ideas in addition in the pretest which follows.

See What You Know And Remember—Pretest 2B

Work these exercises carefully, doing as many problems as you can. Write the answers below each question.

1. 9 +7 ———	**2.** 62 +53 ———	**3.** 38 +23 ———	**4.** 75 + 5 ———	**5.** 275 +351 ———

6. 3,253
+2,926

7. 8
7
6
9
1
4
+6

8. 5,687
+3,445

9. 2,011
+1,989

10. 3,456
2,235
4,197
1,469
+2,106

11. 5,976
2,753
385
4
\+ 56

12. Mr. Carrasco worked overtime for the post office during December. His wages during that time period were $148, $172, $216, and $157. What were his total earnings?

13. Find the sum of 5,897; 290; 1,035; 87; 491; and 34.

14. A committee from work raised money for the Cancer Fund. Each department contributed some money. How much money was donated if the sums collected were $435; $2,861; $3,085; $762; and $1,530?

15. Find the number of feet of fencing needed to enclose a triangular plot of ground whose sides are 73 ft., 69 ft., and 91 ft.

Now turn to the end of the chapter to check your answers. Add up all that you had correct.

A Score of	*Means That You*
14–15	Did very well. You can move to Chapter 3.
12–13	Know this material except for a few points. Read the sections about the ones you missed.
10–11	Need to check carefully on the sections you missed.
0–9	Need to work with this part of the chapter to refresh your memory and improve your skills.

Questions	*Are Covered in Section*
1	2.8
2–6, 8–10	2.9
7	2.10
11, 13	2.11
12, 14	2.13
15	2.14

2.8 Every Number Has Its Place

And every place has a value. You will recall that in Chapter 1 we mentioned that the numeral 23 has 2 tens and 3 ones. Sometimes we find that in adding two one-place addends we get a sum greater than 10. What happens then is shown in the following example, where the 1 is *carried* over into the tens place.

Problem

$$\begin{array}{r|r} & 7 \\ & 6 \\ \hline 1 & 3 \end{array}$$

Another way of looking at it — To see why $7 + 6 = 13$, maybe this approach appeals to you. Think of $7 + 6$ as:

$$7 + 3 + 3 \text{ or}$$
$$10 + 3 = 13$$

Our number system is *base ten* and this method may speed up your work and increase your accuracy.

Add $9 + 7$. Think of it as $9 + 1 + 6$. This gives you $10 + 6 = 16$.

Practice Exercise 18

The following exercise provides you with single addend addition. Remember, practice makes perfect, so work this exercise carefully and try to get all 45 problems correct. Each sum for these problems is 10 or more than 10. Answers to this test and subsequent practice exercises begin on page 59.

1. $\begin{array}{r} 6 \\ +8 \\ \hline \end{array}$	**2.** $\begin{array}{r} 5 \\ +9 \\ \hline \end{array}$	**3.** $\begin{array}{r} 5 \\ +6 \\ \hline \end{array}$	**4.** $\begin{array}{r} 9 \\ +2 \\ \hline \end{array}$	**5.** $\begin{array}{r} 8 \\ +7 \\ \hline \end{array}$
6. $\begin{array}{r} 6 \\ +9 \\ \hline \end{array}$	**7.** $\begin{array}{r} 6 \\ +5 \\ \hline \end{array}$	**8.** $\begin{array}{r} 7 \\ +4 \\ \hline \end{array}$	**9.** $\begin{array}{r} 8 \\ +5 \\ \hline \end{array}$	**10.** $\begin{array}{r} 6 \\ +7 \\ \hline \end{array}$
11. $\begin{array}{r} 3 \\ +9 \\ \hline \end{array}$	**12.** $\begin{array}{r} 9 \\ +9 \\ \hline \end{array}$	**13.** $\begin{array}{r} 7 \\ +6 \\ \hline \end{array}$	**14.** $\begin{array}{r} 4 \\ +8 \\ \hline \end{array}$	**15.** $\begin{array}{r} 9 \\ +5 \\ \hline \end{array}$

16. 9 +4	**17.** 8 +6	**18.** 7 +8	**19.** 8 +9	**20.** 3 +8
21. 9 +7	**22.** 8 +8	**23.** 3 +7	**24.** 4 +6	**25.** 8 +4
26. 9 +6	**27.** 9 +3	**28.** 5 +7	**29.** 9 +1	**30.** 7 +9
31. 4 +7	**32.** 9 +8	**33.** 5 +8	**34.** 2 +8	**35.** 6 +6
36. 5 +5	**37.** 4 +9	**38.** 1 +9	**39.** 8 +3	**40.** 7 +3
41. 7 +7	**42.** 8 +2	**43.** 7 +5	**44.** 2 +9	**45.** 6 +4

2.9 Place Value And Column Addition

If you can work with number combinations and place value, you can do any addition problem. For example: suppose we had to find the sum of 85 and 73.

When the sum of numbers in *any* column, ones-place, tens-place, etc., is greater than 10, we carry to the next column as illustrated below.

Problem	Step 1	Step 2
85 +73	8 5 + 7 3 8	8 5 + 7 3 1 5 8
	Add the *ones* place: 5 + 3 = 8	Add the *tens* place: 8 + 7 = 15 Place the 5 in the *tens* place and carry the 1 to the *hundreds* place.

Thus: 85 + 73 = 158.

Practice Exercise 19

In these examples you are asked to carry to the next column. Be sure you understand the preceding example before you begin. Add.

1. $\begin{array}{r} 63 \\ +61 \\ \hline \end{array}$

2. $\begin{array}{r} 93 \\ +54 \\ \hline \end{array}$

3. $\begin{array}{r} 723 \\ +536 \\ \hline \end{array}$

4. $\begin{array}{r} 803 \\ +392 \\ \hline \end{array}$

5. 423 + 704 =

6. $\begin{array}{r} 9{,}438 \\ +2{,}511 \\ \hline \end{array}$

7. $\begin{array}{r} 85{,}423 \\ +72{,}541 \\ \hline \end{array}$

8. $\begin{array}{r} 923{,}003 \\ +450{,}892 \\ \hline \end{array}$

9. 8,432,000 + 4,527,000 =

10. $\begin{array}{r} 909 \\ +990 \\ \hline \end{array}$

Example

You worked 8 hrs. more than your usual 35 hrs. last week. What was the total number of hours you worked?

Solution

The word *total*, which is one of our clue words, says that we must add. Add: 35 + 8.

Problem	Step 1	Step 2
$\begin{array}{r} 35 \\ +\ 8 \\ \hline \end{array}$	$\begin{array}{r} 1 \\ 35 \\ +\ \ 8 \\ \hline 3 \end{array}$	$\begin{array}{r} 1 \\ 35 \\ +\ \ 8 \\ \hline 43 \end{array}$
	Add: 5 + 8 = 13 Place the 3 in the *ones* column and carry the 1 to the *tens* column.	Add the 1 you carried to the 3 in the *tens* column. 1 + 3 = 4

Thus: 35 + 8 = 43.

Practice Exercise 20

Find the sum for each of these exercises. You are asked to carry into the tens column as in the previous illustration.

1. 79 + 5 = ____
2. 88 + 9 = ____
3. 76 + 8 = ____
4. 78 + 3 = ____
5. 87 + 7 = ____
6. 68 + 9 = ____
7. 58 + 7 = ____
8. 85 + 9 = ____
9. 49 + 6 = ____
10. 39 + 7 = ____
11. 22 + 9 = ____
12. 56 + 6 = ____
13. 47 + 6 = ____
14. 29 + 8 = ____
15. 58 + 4 = ____
16. 48 + 5 = ____
17. 13 + 9 = ____
18. 79 + 9 = ____
19. 57 + 3 = ____
20. 29 + 7 = ____

Numbers can be carried to any column. Look at the two following examples before beginning the next practice exercise.

Problem	Step 1	Step 2	Step 3
453 +284 ———	4 5 [3] + 2 8 [4] ——— [7]	1 4 [5] 3 + 2 [8] 4 ——— [3] 7	[1] [4] 5 3 + [2] 8 4 ——— [7] 3 7
	Add: 3 + 4 = 7	Add: 5 + 8 = 13 Place the 3 in the *tens* place. Carry the 1 to the *hundreds* place.	Add: 4 + 2 + 1 = 7

Thus: 453 + 284 = 737.

Problem	Step 1	Step 2	Step 3	Step 4
5,931 +2,345	5 9 3 1 + 2 3 4 5 6	5 9 3 1 + 2 3 4 5 7 6	1 5 9 3 1 + 2 3 4 5 2 7 6	1 5 9 3 1 + 2 3 4 5 8 2 7 6
	Add: 1 + 5 = 6	Add: 3 + 4 = 7	Add: 9 + 3 = 12 Place the 2 in the *hundreds* place and carry the 1 to the *thousands* place.	Add: 1 + 5 + 2 = 8

Thus:
```
  5,931
 +2,345
 ------
  8,276.
```

Practice Exercise 21

Add these problems.

1. 48 + 33

2. 757 + 272

3. 5,432 + 3,707

4. 39,000 + 19,873

5. 459,321 + 190,005

6. 67 + 58

7. 296 + 130

8. 6,737 + 2,922

9. $23,450 + 18,320

10. 3,450,000 + 2,720,000

Sometimes it is necessary to carry in more than one column. Look at this example:

Example

Shop A produced 148 articles yesterday while Shop B made 156 of the same articles. How many did they make altogether?

Solution

The words *how many* indicate that you must add because you have two *parts* of the total, 148 and 156.

Problem	Step 1	Step 2	Step 3
148 +156	1 1 4 8 + 1 5 6 4	1 1 1 4 8 + 1 5 6 0 4	1 1 4 8 + 1 5 6 3 0 4
	Add: 8 + 6 = 14 Carry the 1.	Add: 1 + 4 + 5 = 10 Carry the 1.	Add: 1 + 1 + 1 = 3

Thus: 148 + 156 = 304.

Practice Exercise 22

Add the following numbers.

1. 98 +56
2. 732 +481
3. 3,932 +1,587
4. 53,493 +37,507
5. 573,258 +273,852
6. 77 +66
7. 856 +375
8. 4,803 +5,219
9. 33,337 +27,082
10. 7,832,000 +5,989,000

2.10 Longer Columns Of Numbers

Adding longer columns of numbers sometimes results in carrying more than a one into the next column. Look at this problem:

Problem

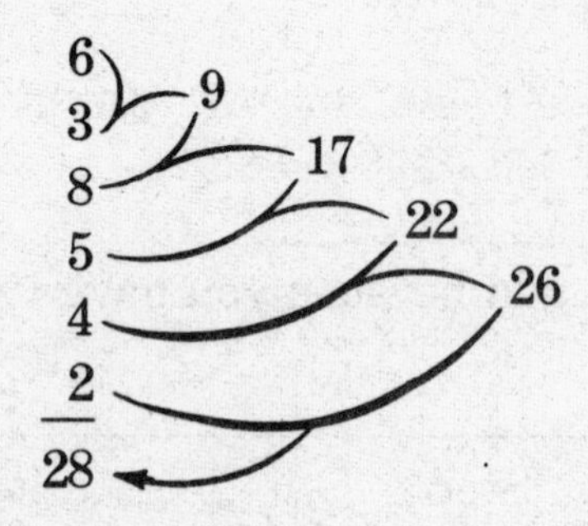

Step 1: $6 + 3 = 9$
Step 2: $9 + 8 = 17$
Step 3: $17 + 5 = 22$
Step 4: $22 + 4 = 26$
Step 5: $26 + 2 = 28$

Practice Exercise 23

Find the sum of each of the following problems.

1.	2.	3.	4.	5.
2	9	4	1	9
2	7	1	8	9
5	6	9	0	7
+8	+2	+1	+3	+6

6.	7.	8.	9.	10.
8	3	1	5	7
8	8	1	9	8
2	5	7	1	4
4	4	1	3	4
+3	+2	+1	+3	+5

2.11 When The Columns Are Broken

Up to the present moment, you have added groups of numbers that contain the same number of digits in each addend. What happens if you are asked to find the sum of a group of addends like 56 + 137 + 9 + 7,638 + 589? Write each addend so that the numbers with the same place value lie in the same column. The 6 in 56, the 7 in 137, the 9, the 8 in 7,638, and the 9 in 589 are all numbers in the ones place, so they are lined up in column form like this:

```
      5 6
    1 3 7
        9
7 , 6 3 8
    5 8 9
```

This is called *broken-column addition*, and once the numbers are arranged correctly, the procedure for adding the numbers is the same as before.

Problem	Step 1	Step 2	Step 3	Step 4
56 137 9 7,638 + 589	3 5 6 1 3 7 9 7 6 3 8 + 5 8 9 9	2 3 5 6 1 3 7 9 7 6 3 8 + 5 8 9 2 9	1 2 5 6 1 3 7 9 7 6 3 8 + 5 8 9 4 2 9	1 5 6 1 3 7 9 7 6 3 8 + 5 8 9 8 4 2 9
	Add: 6 + 7 + 9 + 8 + 9 = 39 Write the 9 and carry 3.	Add: 3 + 5 + 3 + 3 + 8 = 22 Write the 2 and carry 2.	Add: 2 + 1 + 6 + 5 = 14 Write the 4 carry 1.	Add: 1 + 7 = 8

Thus: 56 + 137 + 9 + 7,638 + 589 = 8,429.

Practice Exercise 24

Find the sum of each of the following problems.

1.
```
     70
      5
  8,769
+   127
```

2.
```
    635
     28
  1,038
+    77
```

3.
```
     45
    399
+10,037
```

4.
```
      8
    125
  3,456
+    29
```

5.
```
      4
  3,456
    907
 10,203
+    56
```

6.
```
      7
     31
  3,987
     53
+     7
```

7.
```
     94
    958
      9
+ 6,738
```

8.
```
  7,756
  2,398
      5
    420
+ 1,006
```

In questions 9 and 10, rewrite the problem in column form before adding.

9. Find the sum of 3,257; 4,009; 138; 4; 67; and 3,470.

10. Add 1; 1,002; 208; 4,001; 20,009.

2.12 Know The Combinations

You can see that skill and speed in addition depend very much on knowing how numbers combine. There are basic number combinations using all the symbols from 0 to 9. These are well worth practicing to gain skill because they save so much time when you know them.

Any number can be added to itself or to any of the other numbers. For example, a problem can be 0 + 0, or 0 + each of the other nine numbers. Or it can be 1 + 0, or 1 + each of the other nine numbers. Thus there are *ten* combinations for each number, or one hundred altogether. However, once you know these hundred combinations you can unlock any addition problem.

Practice Exercise 25

This time we suggest that you do not write the answers in the book. Write your answers on a separate piece of paper so you can use this exercise for practice as many times as you wish. Don't write the problem. Place your paper under the first row and write each answer. Then fold back the paper and place it under the next row. Write the answers, fold back the paper, and so forth, until you have all twenty rows. The idea is to see how you can improve by quickly recognizing the answers. Keep your own record on this.

This group of examples is a review. Repeat them as often as you need so they become second nature to you. Remember, practice makes perfect.

1. 0 +1	**2.** 3 +3	**3.** 4 +2	**4.** 7 +0	**5.** 8 +5
6. 1 +1	**7.** 6 +2	**8.** 5 +6	**9.** 9 +9	**10.** 2 +2
11. 1 +5	**12.** 4 +4	**13.** 3 +0	**14.** 0 +7	**15.** 9 +0

16. 5 +0

17. 7 +2

18. 8 +6

19. 2 +7

20. 6 +7

21. 3 +4

22. 0 +2

23. 5 +8

24. 1 +2

25. 4 +6

26. 6 +3

27. 7 +4

28. 2 +1

29. 8 +4

30. 9 +8

31. 4 +8

32. 5 +1

33. 1 +3

34. 3 +4

35. 6 +8

36. 2 +8

37. 0 +0

38. 8 +7

39. 7 +6

40. 9 +1

41. 6 +4

42. 9 +7

43. 8 +3

44. 5 +7

45. 0 +8

46. 3 +1

47. 2 +6

48. 1 +6

49. 4 +3

50. 7 +8

51. 5 +5

52. 1 +4

53. 9 +2

54. 8 +8

55. 2 +3

56. 7 +9

57. 3 +5

58. 4 +5

59. 6 +9

60. 0 +6

61. 8 +2

62. 0 +3

63. 9 +6

64. 2 +0

65. 1 +9

66. 6 +0

67. 5 +9

68. 3 +8

69. 7 +7

70. 4 +7

71. $\begin{array}{r}7\\+5\\\hline\end{array}$	**72.** $\begin{array}{r}4\\+0\\\hline\end{array}$	**73.** $\begin{array}{r}2\\+5\\\hline\end{array}$	**74.** $\begin{array}{r}6\\+6\\\hline\end{array}$	**75.** $\begin{array}{r}3\\+7\\\hline\end{array}$
76. $\begin{array}{r}0\\+9\\\hline\end{array}$	**77.** $\begin{array}{r}1\\+8\\\hline\end{array}$	**78.** $\begin{array}{r}8\\+9\\\hline\end{array}$	**79.** $\begin{array}{r}5\\+2\\\hline\end{array}$	**80.** $\begin{array}{r}9\\+3\\\hline\end{array}$
81. $\begin{array}{r}8\\+0\\\hline\end{array}$	**82.** $\begin{array}{r}2\\+9\\\hline\end{array}$	**83.** $\begin{array}{r}6\\+1\\\hline\end{array}$	**84.** $\begin{array}{r}5\\+4\\\hline\end{array}$	**85.** $\begin{array}{r}7\\+3\\\hline\end{array}$
86. $\begin{array}{r}4\\+9\\\hline\end{array}$	**87.** $\begin{array}{r}9\\+5\\\hline\end{array}$	**88.** $\begin{array}{r}3\\+9\\\hline\end{array}$	**89.** $\begin{array}{r}0\\+5\\\hline\end{array}$	**90.** $\begin{array}{r}1\\+7\\\hline\end{array}$
91. $\begin{array}{r}2\\+4\\\hline\end{array}$	**92.** $\begin{array}{r}4\\+1\\\hline\end{array}$	**93.** $\begin{array}{r}0\\+4\\\hline\end{array}$	**94.** $\begin{array}{r}9\\+4\\\hline\end{array}$	**95.** $\begin{array}{r}3\\+2\\\hline\end{array}$
96. $\begin{array}{r}6\\+5\\\hline\end{array}$	**97.** $\begin{array}{r}7\\+1\\\hline\end{array}$	**98.** $\begin{array}{r}8\\+1\\\hline\end{array}$	**99.** $\begin{array}{r}1\\+0\\\hline\end{array}$	**100.** $\begin{array}{r}5\\+3\\\hline\end{array}$

2.13 More Difficult Word Problems Requiring Addition

Example

The Cordero family are members of a traveling circus. They travel quite a bit. Last summer the family started in Buffalo, New York, played in many cities, and finished in Atlanta, Georgia. This is their path.

From	To	Mileage
Buffalo, N.Y.	Pittsburgh, Pa.	217
Pittsburgh, Pa.	Cincinnati, Ohio	284
Cincinnati, Ohio	Des Moines, Iowa	576
Des Moines, Iowa	St. Paul, Minn.	253
St. Paul, Minn.	Milwaukee, Wis.	330
Milwaukee, Wis.	Omaha, Nebr.	491
Omaha, Nebr.	Tulsa, Okla.	385
Tulsa, Okla.	Houston, Tex.	510
Houston, Tex.	Dallas, Tex.	245
Dallas, Tex.	Memphis, Tenn.	472
Memphis, Tenn.	Louisville, Ky.	389
Louisville, Ky.	Atlanta, Ga.	432

Mr. Cordero wanted to know the total number of miles his automobile would travel on this trip. He had to find the *total* mileage. To find the total, he had to add that long list of numbers. Can you help him?

Solution

```
       6 4
Add:  2 1 7
      2 8 4
      5 7 6
      2 5 3
      3 3 0
      4 9 1
      3 8 5
      5 1 0
      2 4 5
      4 7 2
      3 8 9
      4 3 2
    -------
    4 , 5 8 4  mi.
```

Step 1: The sum of the ones column is 44; carry the 4.

Step 2: The sum of the tens column is 68; carry the 6.

Step 3: The sum of the hundreds column is 45.

That's right; the Cordero family traveled many miles across this land to work. Most people don't travel half that amount going to and from work. Do you have any idea how far you travel each year going to and from work?

Did you notice that the key word *total* appeared in this problem, too? Recall that this word sometimes indicates that you must add in order to find the answer to the problem. Go back to Section 2.7 to review the key words that indicate addition.

Practice Exercise 26

This exercise gives you an opportunity to practice solving word problems along with the basic position facts. Key words in two examples have been underlined for you.

1. If you purchased 6 bottles of soda yesterday and 7 more today, how many bottles have you purchased in the last two days?

2. In five games of cards, Daniel scored the following number of points: 2, 6, 9, 1, and 5. What was his total number of points scored?

3. During a six-week period, your salary checks vary each payday. How much would you earn altogether if your checks were $115, $143, $126, $119, $137, and $125?

4. In a one-week period, an automobile company produced 4,325 sedans, 2,786 station wagons, and 1,938 trucks. How many vehicles were produced?

5. Your street association collects money from 15 families to purchase trees. If the amounts collected are $2, $5, $5, $3, $2, $1, $5, $7, $10, $5, $3, $2, $2, $5, and $1, what is the total amount of money collected?

6. The A & M Auto Company rented 35 cars more this week than the 178 they rented last week. What was the total number of cars rented this week?

7. Cigarette sales last week were 3,423 cartons. This week they increased by 789 cartons. What was the total number of cartons of cigarettes sold this week?

8. Your electric bill states the number of kilowatt hours of electricity you use each month. If you use 232 kilowatt hours in January, 246 in February, 207 in March, 198 in April, and 163 in May, what is the total number of kilowatt hours of electricity used during this five-month period?

9. If you study 3 hrs. today, 2 more tomorrow, 4 on Saturday, and 5 on Sunday, how many hours will you spend studying mathematics this week?

10. The A.I.M. Medical Clinic saw 162 patients on Monday, 146 on Tuesday, 87 on Wednesday, 243 on Thursday, and 109 on Friday. How many patients visited the clinic this week?

2.14 More Problems With Addition

Mathematics has many branches. One of these is geometry. All around us are various geometric shapes. Many of them are familiar to you. Some are rectangles, triangles, and circles. Geometry deals with these shapes, as well as with the measurement of lines, angles, and surfaces. Surely there have been times when you used a ruler and measured the length of lines. Once the measurements were made, you might have added them as illustrated in the following problem.

Example

Grace Badilla wanted to make a vegetable garden in the shape of a rectangle in the rear of her house. It could be 21 ft. long and 13 ft. wide. To protect the vegetables she wanted to enclose the garden with a wire fence. How many feet of fencing did she require?

Solution

We'll begin this problem with a diagram of the garden:

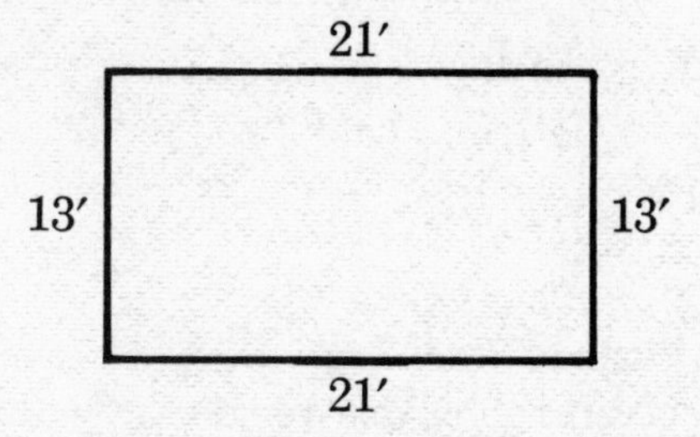

The rectangle, as you recall, has four sides, and each pair of opposite sides are the same length. Thus there are two lengths of 21 ft. and two widths of 13 ft. Since the fencing will be placed around the entire edge of the rectangle, the problem is simply to find the sum:

$$21\text{ ft.} + 21\text{ ft.} + 13\text{ ft.} + 13\text{ ft.}$$

The sum of all the sides of a rectangle or geometric shape is given a special name. It is called the *perimeter*. We can write this word in a simplified form by using the letter P. We can also use an abbreviation (′) for *feet*.

The problem in its simplest form now becomes:

$$P = 21' + 21' + 13' + 13'$$

Finding the sum of the four addends, we get:

$$P = 68'$$

Grace now knew she had to purchase at least 68 ft. of fencing to enclose her garden.

The layout or format of the problem may cause you to wonder why you can't just add all the sides. You can, but you will want to learn a new idea which will be very helpful later when you study more advanced mathematics.

$P = 21' + 21' + 13' + 13'$ is called a *statement of equality* or an *equation.* It is a good idea to learn this form because you will use it often.

Perimeters of other geometric shapes can also be found. Here is an illustration of the triangle:

Example

Find the perimeter of the triangle drawn below. The symbol (″) means inches.

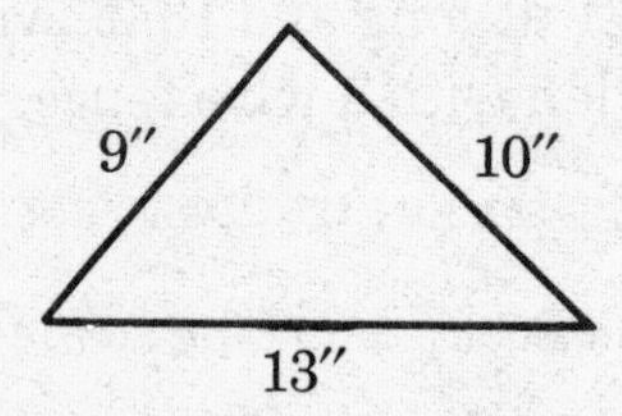

Solution

Perimeter = 9 in. + 10 in. + 13 in.

or

$P = 9'' + 10'' + 13''$

$P = 32''$

Practice Exercise 27

In each of the following, first write the perimeter in the form of an equation, as illustrated in (a) of the first problem. Then solve the problem.

1. Find the perimeter for each of these:

a.

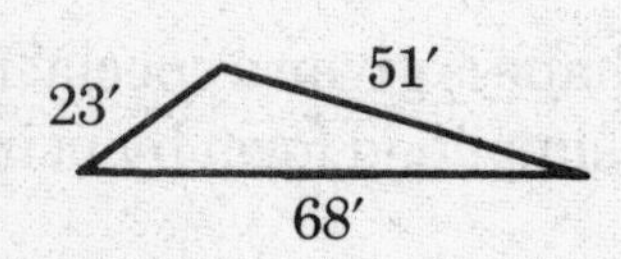

$P = 23' + 51' + 68'$

$P =$

b.

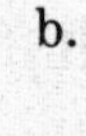

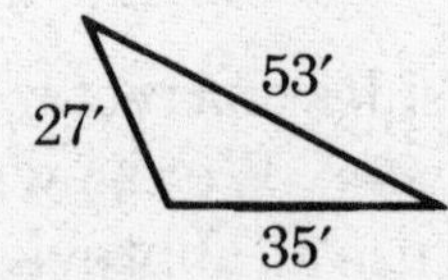

c.

d.

2. Find the amount of fencing required to enclose a *triangular* plot of ground measuring 27 ft., 54 ft., and 69 ft.

   a. Draw a diagram.

   b. Find the perimeter.

3. How many feet of fencing are required to enclose a *rectangular*-shaped garden whose dimensions are 47 ft. by 35 ft.?

Terms You Should Remember

Place value The location of the symbol in the numeral signifying its value.

Triangle A three-sided geometric figure.

Perimeter The sum of all the sides of a geometric figure.

Review of Important Ideas

The tens number in a sum greater than 10 must be carried to the next column on its left.

Geometry is a branch of mathematics dealing with shapes and with measurements of lines, angles, and surfaces.

Rectangles (□) and triangles (◺) are just two of the many shapes.

The perimeter is found by adding *all* the sides of the geometric figure.

An equation is a statement of equality of two quantities. The sign = is placed between them.

Let's Check How Well You're Doing

This half of the chapter completed all ideas associated with adding whole numbers. Most of this material was probably familiar to you. The posttest lets you check on your understanding. Try these examples.

Posttest 2B

Write your answers below each question.

1.
```
  8
 +5
 ---
```

2.
```
  73
 +41
 ---
```

3.
```
  48
 +27
 ---
```

4.
```
  65
 + 5
 ---
```

5.
```
  382
 +146
 ----
```

6.
```
  4,813
 +2,625
 ------
```

7.
```
  9
  5
  3
  3
  6
  8
 +7
 --
```

8.
```
  4,956
 +3,687
 ------
```

9.
```
  4,099
 +2,901
 ------
```

10.
```
  2,624
  3,540
  4,748
  2,794
 +1,960
 ------
```

11.
```
  9,754
  5,860
     17
  1,063
 +  297
 ------
```

12. The first seven people who purchase rolls in the F & R Bakery store on Tuesday morning buy 6, 8, 12, 4, 8, 5, and 18 rolls. What is the total number of rolls purchased?

13. Find the sum of 1,089; 2; 674; 80; and 6,872.

14. During seven workdays you produced 286, 340, 197, 432, 263, 307, and 169 parts per day. How many parts did you complete at the end of that time?

15. Mr. O'Connor wanted to install molding on the ceiling of a 13 ft. by 8 ft. rectangular-shaped room. How many feet of molding are required?

ANSWERS AND EXPLANATIONS TO POSTTEST 2B

1. 13
2. 114
3. 75
4. 70
5. 528
6. 7,438
7. 41
8. 8,643
9. 7,000
10. 15,666
11. 16,991
12. $6 + 8 + 12 + 4 + 8 + 5 + 18 = 61$
13. $1{,}089 + 2 + 674 + 80 + 6{,}872 = 8{,}717$
14. $286 + 340 + 197 + 432 + 263 + 307 + 169 = 1{,}994$
15. P = 13 ft. + 8 ft. + 13 ft. + 8 ft.
P = 42 feet

A Score of	*Means That You*
14–15	Did very well. You can move to Chapter 3.
12–13	Know this material except for a few points. Reread the sections about the ones you missed.
10–11	Need to check carefully on the sections you missed.
0–9	Need to review this part of the chapter again to refresh your memory and improve your skills.

Questions	*Are Covered in Section*
1	2.8
2–6, 8–10	2.9
7	2.10
11, 13	2.11
12, 14	2.13
15	2.14

Hold It!

Now that you have had lots of practice in addition, go back to Section 2.12 and see if you can improve your score on "Know the Combinations." Can you make a perfect score?

Puzzle Time

Try your hand with this cross-number puzzle which reviews the *addition* of *whole numbers*. Do it as you would a regular crossword puzzle. The answers down and across must all check. Two answers, 2 Across and 1 Down, have been filled in for you. The solution appears on page 61.

	1 4			2 1	3 7			4	
5	3		6				7		
8	5		9			10			
11	8					12		13	
		14	15						
	16				17				
18		19						20	21
		22		23		24		25	
26	27			28				29	
				30					

ACROSS

2. 9 + 8 =
5. 6 + 7 =
6. 59 + 44 =
7. 39 + 98 =
8. 8 + 7 =
9. 20,373 + 4,842 =
11. 93 + 5 =
12. 683 + 752 =
14. 56 + 89 + 67 =
17. 107 + 343 =
18. 752 + 398 =
20. 69 + 8 =
22. 872 + 759 + 10,832 =
25. 74 + 7 =
26. 548 + 197 =
28. 97 + 734 + 9 =
29. 37 + 9 =
30. 65 + 6 =

DOWN

1. 1,686 + 2,672 =
2. 5,872 + 2,762 + 1,868 =
3. 497 + 235 =
4. 79 + 4 =
5. 98 + 21 =
6. 8 + 4 =
7. 75 + 79 =
10. 827 + 308 =
13. 25 + 8 =
15. 957 + 745 =
16. 17 + 4 =
17. 23,750 + 20,425 + 1,466 =
19. 345 + 170 =
20. 197 + 4,862 + 2,784 =
21. 47 + 82 + 545 + 42 =
23. 239 + 229 + 19 =
24. 26 + 4 =
27. 28 + 19 =

ANSWERS FOR CHAPTER 2—B

PRETEST 2B

1. 16	2. 115	3. 61	4. 80	5. 626
6. 6,179	7. 41	8. 9,132	**9.** 4,000	**10.** 13,463
11. 9,174	**12.** $693	**13.** 7,834	**14.** $8,673	**15.** 233 ft.

PRACTICE EXERCISE 18

1. 14	**2.** 14	3. 11	**4.** 11	5. 15
6. 15	7. 11	8. 11	**9.** 13	**10.** 13
11. 12	**12.** 18	**13.** 13	**14.** 12	**15.** 14
16. 13	**17.** 14	**18.** 15	**19.** 17	**20.** 11
21. 16	**22.** 16	**23.** 10	**24.** 10	**25.** 12
26. 15	**27.** 12	**28.** 12	**29.** 10	**30.** 16
31. 11	**32.** 17	**33.** 13	**34.** 10	**35.** 12
36. 10	**37.** 13	**38.** 10	**39.** 11	**40.** 10
41. 14	**42.** 10	**43.** 12	**44.** 11	**45.** 10

PRACTICE EXERCISE 19

1. 124	2. 147	3. 1,259	4. 1,195	5. 1,127
6. 11,949	7. 157,964	8. 1,373,895	**9.** 12,959,000	**10.** 1,899

PRACTICE EXERCISE 20

1. 84	2. 97	3. 84	4. 81	5. 94
6. 77	7. 65	8. 94	**9.** 55	**10.** 46
11. 31	**12.** 62	**13.** 53	**14.** 37	**15.** 62
16. 53	**17.** 22	**18.** 88	**19.** 60	**20.** 36

PRACTICE EXERCISE 21

1. 71	**2.** 1,029	**3.** 9,139	**4.** 58,873	**5.** 649,326
6. 125	**7.** 426	**8.** 9,659	**9.** $41,770	**10.** 6,170,000

PRACTICE EXERCISE 22

1. 154	**2.** 1,213	**3.** 5,519	**4.** 91,000	**5.** 847,110
6. 143	**7.** 1,231	**8.** 10,022	**9.** 60,419	**10.** 13,821,000

PRACTICE EXERCISE 23

1. 17	**2.** 24	**3.** 15	**4.** 12	**5.** 31
6. 25	**7.** 22	**8.** 11	**9.** 21	**10.** 28

PRACTICE EXERCISE 24

1. 8,971	**2.** 1,778	**3.** 10,481	**4.** 3,618	**5.** 14,626
6. 4,085	**7.** 7,799	**8.** 11,585	**9.** 10,945	**10.** 25,221

PRACTICE EXERCISE 25

1. 1	**2.** 6	**3.** 6	**4.** 7	**5.** 13
6. 2	**7.** 8	**8.** 11	**9.** 18	**10.** 4
11. 6	**12.** 8	**13.** 3	**14.** 7	**15.** 9
16. 5	**17.** 9	**18.** 14	**19.** 9	**20.** 13
21. 7	**22.** 2	**23.** 13	**24.** 3	**25.** 10
26. 9	**27.** 11	**28.** 3	**29.** 12	**30.** 17
31. 12	**32.** 6	**33.** 4	**34.** 7	**35.** 14
36. 10	**37.** 0	**38.** 15	**39.** 13	**40.** 10
41. 10	**42.** 16	**43.** 11	**44.** 12	**45.** 8
46. 4	**47.** 8	**48.** 7	**49.** 7	**50.** 15
51. 10	**52.** 5	**53.** 11	**54.** 16	**55.** 5
56. 16	**57.** 8	**58.** 9	**59.** 15	**60.** 6
61. 10	**62.** 3	**63.** 15	**64.** 2	**65.** 10
66. 6	**67.** 14	**68.** 11	**69.** 14	**70.** 11
71. 12	**72.** 4	**73.** 7	**74.** 12	**75.** 10
76. 9	**77.** 9	**78.** 17	**79.** 7	**80.** 12
81. 8	**82.** 11	**83.** 7	**84.** 9	**85.** 10
86. 13	**87.** 14	**88.** 12	**89.** 5	**90.** 8
91. 6	**92.** 5	**93.** 4	**94.** 13	**95.** 5
96. 11	**97.** 8	**98.** 9	**99.** 1	**100.** 8

PRACTICE EXERCISE 26

1.
```
  6
 +7
 --
 13
```

2.
```
  2
  6
  9
  1
 +5
 --
 23
```

3.
```
  $115
   143
   126
   119
   137
 + 125
 -----
  $765
```

4.
```
  4,325
  2,786
 +1,938
 ------
  9,049
```

5. $2 + $5 + $5 + $3 + $2 + $1 + $5 + $7 + $10 + $5 + $3 + $2 + $2 + $5 + $1 = $58

6.
```
  178
 + 35
 ----
  213
```

7.
```
  3,423
 +  789
 ------
  4,212
```

8.
```
  232
  246
  207
  198
 +163
 -----
 1,046
```

9.
```
  3
  2
  4
 +5
 --
 14 hrs.
```

10.
```
  162
  146
   87
  243
 +109
 ----
  747
```

PRACTICE EXERCISE 27

1. a. $P = 142'$ b. $P = 115'$ c. $P = 202'$ d. $P = 222''$

2. a.

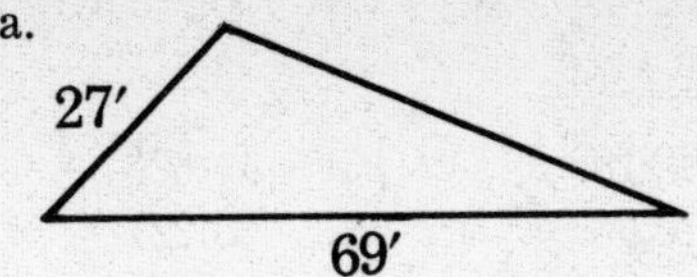

b. $P = 150'$

3. $P = 164'$

CROSS-NUMBER PUZZLE SOLUTION

	(1) 4			(2) 1	(3) 7			(4) 8	
(5) 1	3	■	(6) 1	0	3	■	(7) 1	3	7
(8) 1	5	■	(9) 2	5	2	(10) 1	5	■	■
(11) 9	8	■	■	0	■	(12) 1	4	(13) 3	5
	■	(14) 2	(15) 1	2	■	3	■	3	
	(16) 2	■	7	■	(17) 4	5	0	■	
(18) 1	1	(19) 5	0	■	5	■	■	(20) 7	(21) 7
■	■	(22) 1	2	(23) 4	6	(24) 3	■	(25) 8	1
(26) 7	(27) 4		■	(28) 8	4	0	■	(29) 4	6
	7			(30) 7	1			3	

Good! You are doing well. You are now ready to move on to Chapter 3 and to the pretest.

Subtraction is the second basic operation you need in working with numbers. Finding the difference between two amounts is a subtraction problem. If one can of beans costs 69¢ and another costs 58¢, you want to know the *difference* or how much you would save with the second. If your total purchases cost $3.37 and you have $5.00, you subtract to find the change.

You will see as you work with subtraction that it is not a new idea. In fact, subtraction is the opposite or the *inverse* of addition. The main ideas of subtraction are covered in Pretest 3.

See What You Know And Remember — Pretest 3

Show how well you can do with these examples. Write your answers below each question.

1. $\begin{array}{r} 8 \\ -5 \\ \hline \end{array}$ **2.** $\begin{array}{r} 73 \\ -51 \\ \hline \end{array}$ **3.** $\begin{array}{r} 256 \\ -\ 34 \\ \hline \end{array}$ **4.** $\begin{array}{r} 456 \\ -255 \\ \hline \end{array}$ **5.** $\begin{array}{r} 366 \\ -150 \\ \hline \end{array}$

6. $\begin{array}{r} 96 \\ -\ 3 \\ \hline \end{array}$

7. $\begin{array}{r} 15 \\ -\ 9 \\ \hline \end{array}$

8. $\begin{array}{r} 73 \\ -\ 8 \\ \hline \end{array}$

9. $\begin{array}{r} 325 \\ -171 \\ \hline \end{array}$

10. $\begin{array}{r} 812 \\ -567 \\ \hline \end{array}$

11. $\begin{array}{r} 3{,}069 \\ -2{,}154 \\ \hline \end{array}$

12. 2,001 − 1,983 =

13. $\begin{array}{r} 4{,}000 \\ -\ \ 208 \\ \hline \end{array}$

14. What is the difference between 58¢ and 69¢?

15. On the first day of a two-day charity drive $42,607 was collected. If, on the second day, $6,118 less than the first day was collected, how much was collected the second day?

16. How much change do you receive from a ten dollar bill after a purchase of 4 dollars 82 cents?

17. How much less is 212 than 632?

18. The value of a house decreased $1,150 from its original purchase price of $19,345. What is its present value?

19. If two sides of a triangle measure 74 ft. and 58 ft., what is the length of the third side if the perimeter of the triangle is 191 ft.?

20. Show the result of subtracting 5 from 8 on the number line below.

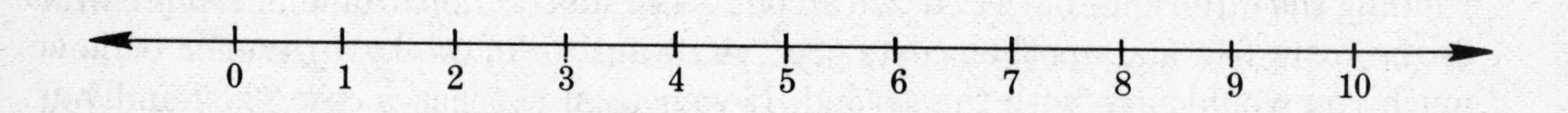

Check your answers by turning to the end of the chapter. Add all that you had correct.

A Score of	*Means That You*
18–20	Did very well. You can move to Chapter 4.
16–17	Know the material except for a few points. Read the sections about the ones you missed.
13–15	Need to check carefully on the sections you missed.
0–12	Need to work with this chapter to refresh your memory and improve your skills.

Questions	*Are Covered in Section*
1, 2, 4, 5	3.2
20	3.3
3, 6	3.4
14, 17	3.5
16	3.6
7	3.7
8–13	3.8
15, 18, 19	3.9

3.1 What Happens In Subtraction?

If you have 8 coins and I "take away" 5 of them, what do you have left? This is a problem in subtraction.

You have 8 coins.

I "take away" 5 coins.

You have 3 coins left.

You write the number that you "take away" with a minus sign to its left, under the amount you had when you started. Taking away is another name for subtraction and the minus sign (−) is used to indicate the operation subtraction. Thus:

$$\begin{array}{r} 8 \text{ coins} \\ -5 \text{ coins} \\ \hline 3 \text{ coins} \end{array}$$

Subtraction is not a new operation for you. It is the inverse of addition. Webster defines *inverse* as the *direct opposite,* and subtraction is the opposite operation of addition. Instead of *adding on* we are *taking away.*

What does this mean? If we subtract 5 from 8, we write it vertically like this:

$$\begin{array}{r} 8 \\ -5 \\ \hline ? \end{array}$$

or horizontally like this: $8 - 5 = ?$ Since subtraction is the opposite of addition, we mean $5 + ? = 8$. The examples:

$$\begin{array}{r} 8 \\ -5 \\ \hline ? \end{array} \quad \text{or} \quad \begin{array}{r} 5 \\ +? \\ \hline 8 \end{array}$$

mean the same thing. The missing number is 3:

$$\begin{array}{r} 8 \\ -5 \\ \hline 3 \end{array} \quad \text{or} \quad \begin{array}{r} 5 \\ +3 \\ \hline 8 \end{array}$$

3.2 The Language Of Subtraction

Example

Mrs. Maxson paid $8 to the Day Care Center last week and only $5 this week to pay for the care of her daughter. How much less did it cost her this week than last week?

Solution

To find out how much less the care cost, you subtract $5 from $8. There are names for the parts of a subtraction example: the minuend, the subtrahend, and the remainder or difference.

Written in column form, the problem would look like this:

	8	minuend
The *minus* sign indicates subtraction ↗	−5	subtrahend
	3	difference or remainder

You notice that subtraction also has a language. These are terms that you must know to understand subtraction.

Terms You Should Remember

Subtract To take away.

Minuend That number from which another number is to be subtracted.

Subtrahend The number to be subtracted from another.

Minus The sign (−) of subtraction.

Difference or *Remainder* That which is left after a part has been taken away.

You can *check* your answer in subtraction by adding the answer (remainder) to the number you subtracted (subtrahend) and the result should be the number you are subtracting from (minuend).

Example	*Check*	
8	5	subtrahend
−5	+3	remainder
3	8	minuend

Before you start the practice exercise, let's discuss a few basic facts you are already aware of.

You know that zero taken away from a number will make no change in the number.

$$\begin{array}{r}4\\-0\\\hline 4\end{array}\qquad\begin{array}{r}7\\-0\\\hline 7\end{array}\qquad\begin{array}{r}9\\-0\\\hline 9\end{array}$$

Similarly, *one* taken away from a number will leave the number *one* lower.

$$\begin{array}{r}6\\-1\\\hline 5\end{array}\qquad\begin{array}{r}8\\-1\\\hline 7\end{array}\qquad\begin{array}{r}3\\-1\\\hline 2\end{array}$$

Subtracting the same quantity from itself will always result in a remainder of 0.

$$\begin{array}{r}9\\-9\\\hline 0\end{array}\qquad\begin{array}{r}6\\-6\\\hline 0\end{array}$$

You can also figure quickly examples taking away two, three, etc. The practice exercise that follows has all possible combinations of single-digit examples. Write your answers on a separate piece of paper so you can use this exercise for practice as many times as you wish. Don't write the problem. Place your paper under the first row and write each answer. Then fold back the paper and place it under the next row. Write the answers, fold back the paper, and so forth, until you have completed the exercise. The idea is to see how you can improve by recognizing the answers more and more quickly. Keep your own record on this.

Practice Exercise 28

Subtract and check by adding. Try to get all correct.

1. 6 −1	**2.** 4 −4	**3.** 5 −0	**4.** 8 −2	**5.** 8 −3
6. 1 −1	**7.** 6 −4	**8.** 3 −2	**9.** 6 −5	**10.** 0 −0
11. 4 −3	**12.** 9 −2	**13.** 1 −0	**14.** 7 −4	**15.** 7 −6
16. 7 −1	**17.** 7 −7	**18.** 6 −0	**19.** 9 −8	**20.** 9 −9
21. 2 −1	**22.** 7 −5	**23.** 8 −4	**24.** 4 −2	**25.** 7 −0
26. 3 −1	**27.** 8 −5	**28.** 2 −0	**29.** 5 −3	**30.** 8 −8
31. 8 −7	**32.** 7 −2	**33.** 6 −6	**34.** 3 −0	**35.** 6 −3

36. 9 − 5 = ___

37. 8 − 0 = ___

38. 9 − 6 = ___

39. 9 − 4 = ___

40. 4 − 1 = ___

41. 9 − 3 = ___

42. 9 − 7 = ___

43. 5 − 2 = ___

44. 8 − 1 = ___

This last group of examples is written horizontally. Write each of them in column form before subtraction.

45. 3 − 3 = 46. 9 − 0 = 47. 6 − 2 = 48. 9 − 1 =

49. 5 − 5 = 50. 5 − 1 = 51. 8 − 6 = 52. 7 − 3 =

53. 5 − 4 = 54. 4 − 0 = 55. 2 − 2 =

Example

Air fare from Provincetown or Boston, Massachusetts, to Washington, D.C., is shown in the table. How much *less* is it to travel to Washington, D.C., from Boston than from Provincetown?

	Air Fare from Provincetown	Air Fare from Boston
Houston	$144	$128
New Orleans	$128	$112
Miami or Fort Lauderdale	$122	$106
West Palm Beach	$118	$102
Orlando or Tampa	$118	$102
Atlanta	$101	$ 85
Washington, D.C.	$ 59	$ 43
Baltimore	$ 57	$ 41
Philadelphia	$ 52	$ 36

Solution

To find the difference in air fares we must subtract the two costs in air fare. Subtract $43 from $59 like this:

Problem	Step 1	Step 2	Check
$59 − 43	$ 5 [9] − 4 [3] [6]	$ [5] 9 − [4] 3 $ [1] 6	$43 + 16 $59
	Subtract the *ones* column first: 9 − 3 = 6	Subtract the *tens* column: 5 − 4 = 1	

Thus: The difference of $59 − $43 = $16.

Practice Exercise 29

Subtract the following examples. Check by adding.

1. 35 −23	**2.** 47 −12	**3.** 25 −14	**4.** 55 −23	**5.** 74 −11
6. 88 −15	**7.** 36 −13	**8.** 78 −21	**9.** 65 −24	**10.** 56 −31

The same method is used when subtracting numbers with more than two places. You begin subtracting in the ones column and move to the left. Be sure the numbers are written in column form with digits having the same place value under one another.

Subtractions that result in zero are meaningful. Remember $101 is not the same as $11, so be sure you place the zero in the answer when it is needed.

Example

Quarterly union dues of $123 and $112 were deducted from two of Mr. Godwin's checks. By how much was the second deduction decreased?

Solution

Once again we must subtract $112 from $123 to find the decrease.

Problem	Step 1	Step 2	Step 3	Check
\$123 − 112	\$ 1 2 [3] − 1 1 [2] [1]	\$ 1 [2] 3 − 1 [1] 2 [1] 1	\$ [1] 2 3 − [1] 1 2 [0] 1 1	\$112 + 011 \$123
	Subtract the *ones* place: 3 − 2 = 1	Subtract the *tens* place: 2 − 1 = 1	Subtract the *hundreds* place: 1 − 1 = 0	

The difference between \$123 and \$112 = \$011. The zero in this case precedes the whole number and shows there are no hundreds. When zero is the initial digit in a number it is customary not to write it. The result is \$11.

Practice Exercise 30

Subtract. Check by adding. Be sure zero is included in your answer when needed.

1. 68 − 43
2. 629 − 304
3. 2,985 − 1,383
4. 24,679 − 22,569
5. 623,481 − 212,401
6. 39 − 25
7. 843 − 641
8. 3,497 − 3,100
9. 38,437 − 16,202
10. \$4,385,000 − 3,152,000
11. 437 − 212
12. 8,543 − 2,123
13. 63,439 − 63,212
14. \$3,795 − 3,550
15. 847,356 − 615,301

3.3 Help From The Number Line

Subtraction can also be done on the number line. For example, 8 − 5 = ? can be shown as the reverse operation of addition. Draw a line from 0 to 8 units representing the amount you are starting with. From 8, draw a line 5 units long extending in the *reverse* direction. This represents the amount you are subtracting. The answer is the segment extending from 0 to 3 units.

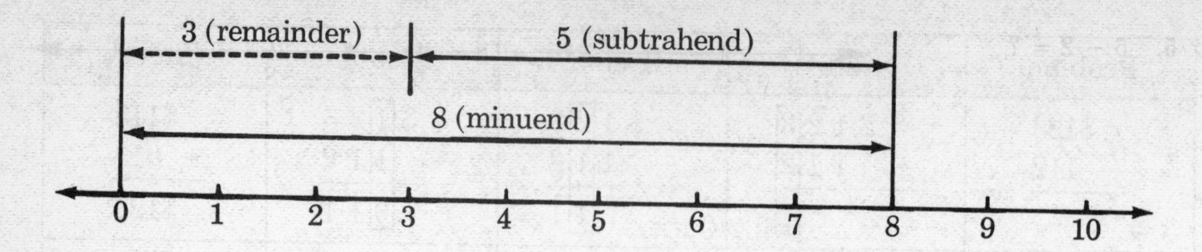

or $8 - 5 = 3$.

Can we subtract a larger number from a smaller one like $5 - 8 = ?$ Represent 5 as a line extending from 0 to 5 units and then reverse the direction of the number you are taking away by drawing a line 8 units to the left. This will extend beyond the zero of the number line:

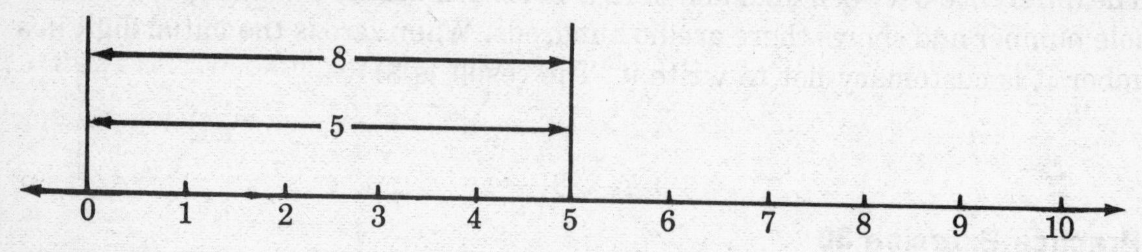

There is no positive number associated with points to the left of zero. You see that an answer to this example is impossible if you use a number line which begins at zero and extends only to the right as far as you desire.

Check to see if you understand this idea by doing the practice exercise which follows.

Practice Exercise 31

Using the given number line for each problem, illustrate the subtraction indicated. Write the problem in column form before you begin.

1. $9 - 3 = ?$

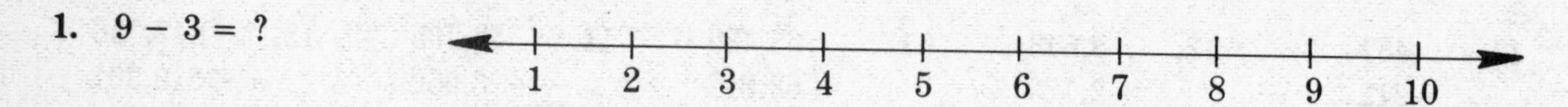

2. $8 - 6 = ?$

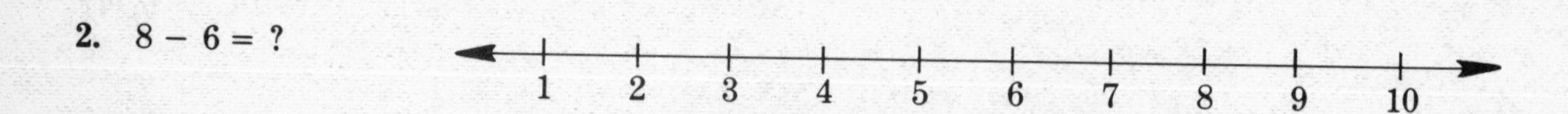

3. $10 - 3 = ?$

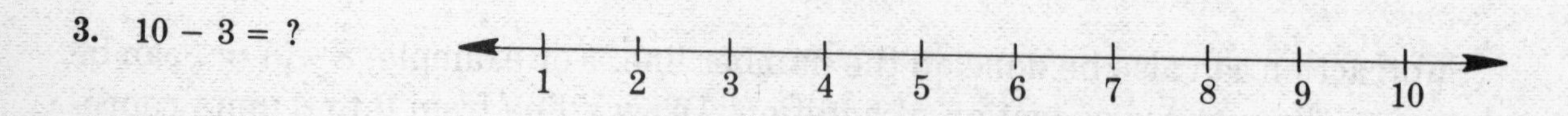

4. $7 - 3 = ?$

1 2 3 4 5 6 7 8 9 10

5. 5 − 2 = ?

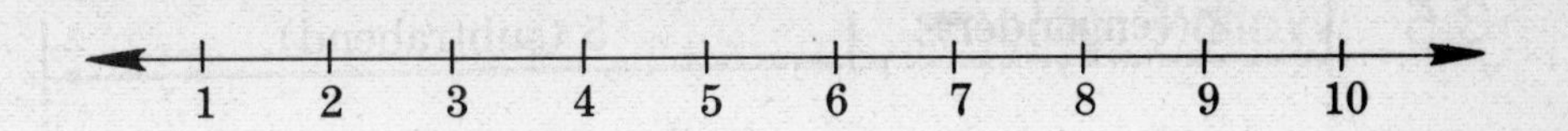

3.4 Unequal Place Subtraction

Example

Edna saved $43 from the selling price of $255 when purchasing a sofa for cash. What was the cost of the sofa?

Solution

You are subtracting a two-place number from a three-place number; $255 − $43 = ?

Problem	Step 1	Step 2	Step 3	Check
$255 − 43	$ 2 5 [5] − 4 [3] [2]	$ 2 [5] 5 − [4] 3 [1] 2	$ [2] 5 5 − [ ] 4 3 $ [2] 1 2	$ 43 + 212 $255
	Subtract: 5 − 3 = 2	Subtract: 5 − 4 = 1	Subtract: 2 − 0 = 2 Consider 43 as 043.	

Thus: $255 − $43 = $212.

Practice Exercise 32

Subtract. Check by adding.

1. 35 − 4
2. 368 − 41
3. 546 − 32
4. 4,332 − 120
5. 6,747 − 43

Write in column form before subtracting.

6. 483 − 12 =
7. 5,439 − 27 =
8. 387 − 12 =
9. 48 − 3 =
10. $432,000 − $12,000 =

3.5 Word Problems

Before you begin, go back to "General Instructions" in Chapter 2, Section 2.7, to review the methods used to solve any word problem.

Key Words For Problems Solved By Subtracting

Certain key words indicate that the problem will be solved by subtracting.

1. *Decreased* — "What is the *decreased* cost of the sofa?"
2. *Less* — "How much *less* is it from Boston than from Provincetown?"
3. *Difference* — "What is the *difference* in price?"
4. *Fewer* — "How many *fewer* jokes did he tell?
5. *Remaining* — "How many people are *remaining* for the weekend?"

Practice Exercise 33

Do these problems, after you search out key words or phrases.

1. You had $8 before you went to the drive-in movie with the family. After spending $5, how many dollars remained?

2. One new camera can be purchased for $73 while another brand costs $51. How much less money does the cheaper camera cost?

3. A drill which usually costs $29 is on sale this week for $21. What is the difference in price?

4. Last week the fishing boats caught 338,400 lb. of fish. This week the poundage dropped to 231,200. How many fewer pounds of fish were caught this week?

5. Mrs. Catablanco earned $8,764 in 1973. She earned only $6,560 in 1974. By how much did her salary decrease?

6. By how much do the houses shown above differ in price?

7. The town baseball team won 31 games this season while playing a total of 45 games. How many games did they lose?

8. The boat *Widgeon* now costs $1,075 but the price will rise to $1,195 next month. How much will you save by buying it now?

9. A pair of earrings sells for $14 while a necklace costs $25. How much more does the necklace cost than the earrings?

10. On Monday 43 cans of vegetables were on the shelf of the L & Z Store. By Wednesday, only 22 cans remained. How many were sold during that time?

3.6 Subtraction By Addition

Subtraction can be performed by thinking of it as a related addition problem. This is especially true when you purchase an item in a store and the salesperson gives you change from your bill. Suppose your purchase amounted to $4.38 and you paid with a $5.00 bill.

Instead of subtracting $4.38 from $5.00, some people would treat this as an addition example. They would keep adding to $4.38 until they got to $5.00. They would say, for example, \$4.39, \$4.40, \$4.50, \$5.00 — meaning that they gave you 1¢, 1¢, 10¢, and then a 50¢ piece, or a total of 62¢ change. This is illustrated below.

4 dollars 38 cents + 1 cent =
4 dollars 39 cents + 1 cent =
4 dollars 40 cents + 10 cents =
4 dollars 50 cents + 50 cents =
5 dollars

This is sometimes a simpler way of subtracting.

Example

How much change would you get from a 10 dollar bill after a purchase of 3 dollars and 59 cents?

Solution

10 dollars − 3 dollars 59 cents = ?

Adding from 3 dollars 59 cents to 10 dollars you get

3 dollars 59 cents + 1 cent =
3 dollars 60 cents + 5 cents =
3 dollars 65 cents + 10 cents =
3 dollars 75 cents + 25 cents =
4 dollars + 1 dollar =
5 dollars + 5 dollars =
10 dollars

Your change is \$5 + \$1 + 25¢ + 10¢ + 5¢ + 1¢ or 6 dollars and 41 cents.

Practice Exercise 34

Subtract by adding. The first one has been done for you.

1. 5 dollars − 3 dollars 74 cents = 1 dollar 26 cents
   __0__ \$5; __1__ \$1; __0__ 50¢; __1__ 25¢; __0__ 10¢; __0__ 5¢; __1__ 1¢

2. 2 dollars − 1 dollar 59 cents =
   ____ \$5; ____ \$1; ____ 50¢; ____ 25¢; ____ 10¢; ____ 5¢; ____ 1¢

3. 10 dollars − 9 dollars 95 cents =
   ____ \$5; ____ \$1; ____ 50¢; ____ 25¢; ____ 10¢; ____ 5¢; ____ 1¢

4. 20 dollars − 11 dollars 43 cents =

_______ $5; _______ $1; _______ 50¢; _______ 25¢; _______ 10¢; _______ 5¢; _______ 1¢

5. 7 dollars − 6 dollars 67 cents =

_______ $5; _______ $1; _______ 50¢; _______ 25¢; _______ 10¢; _______ 5¢; _______ 1¢

3.7 Numbers Can Be Exchanged

Example

You spend $8 from a total of $13. How much do you have left?

Solution

You must subtract $8 from $13 or $13 − 8 = ?

Problem	Step 1	Step 2	Check
$13 − 8	0 1 $ ~~1~~3 − 8	0 1 $ ~~1~~ 3 − 8 5	$ 8 + 5 $13
	In the *ones* column you can't subtract 8 from 3, so *exchange* 1 ten in the minuend for 10 ones. You have 0 tens and 13 ones.	Subtract: 13 − 8 = 5	

Thus: $13 − 8 = 5.

Practice Exercise 35

Subtract these problems. Check by adding. As was suggested in Section 3.2, write your answers on a separate piece of paper, folding back the paper after each row, so that you can review this exercise as many times as you wish. Do not write in the book.

1.	2.	3.	4.	5.
12 − 9	16 − 7	12 − 8	14 − 6	18 − 9

6. $\begin{array}{r} 11 \\ -\ 6 \\ \hline \end{array}$	7. $\begin{array}{r} 12 \\ -\ 7 \\ \hline \end{array}$	8. $\begin{array}{r} 13 \\ -\ 5 \\ \hline \end{array}$	9. $\begin{array}{r} 17 \\ -\ 8 \\ \hline \end{array}$	10. $\begin{array}{r} 11 \\ -\ 8 \\ \hline \end{array}$
11. $\begin{array}{r} 15 \\ -\ 9 \\ \hline \end{array}$	12. $\begin{array}{r} 14 \\ -\ 7 \\ \hline \end{array}$	13. $\begin{array}{r} 10 \\ -\ 5 \\ \hline \end{array}$	14. $\begin{array}{r} 16 \\ -\ 8 \\ \hline \end{array}$	15. $\begin{array}{r} 11 \\ -\ 7 \\ \hline \end{array}$
16. $\begin{array}{r} 14 \\ -\ 5 \\ \hline \end{array}$	17. $\begin{array}{r} 11 \\ -\ 4 \\ \hline \end{array}$	18. $\begin{array}{r} 11 \\ -\ 9 \\ \hline \end{array}$	19. $\begin{array}{r} 10 \\ -\ 7 \\ \hline \end{array}$	20. $\begin{array}{r} 12 \\ -\ 4 \\ \hline \end{array}$
21. $\begin{array}{r} 10 \\ -\ 9 \\ \hline \end{array}$	22. $\begin{array}{r} 10 \\ -\ 6 \\ \hline \end{array}$	23. $\begin{array}{r} 10 \\ -\ 4 \\ \hline \end{array}$	24. $\begin{array}{r} 11 \\ -\ 2 \\ \hline \end{array}$	25. $\begin{array}{r} 10 \\ -\ 3 \\ \hline \end{array}$
26. $\begin{array}{r} 16 \\ -\ 9 \\ \hline \end{array}$	27. $\begin{array}{r} 13 \\ -\ 8 \\ \hline \end{array}$	28. $\begin{array}{r} 15 \\ -\ 6 \\ \hline \end{array}$	29. $\begin{array}{r} 10 \\ -\ 2 \\ \hline \end{array}$	30. $\begin{array}{r} 10 \\ -\ 8 \\ \hline \end{array}$
31. $\begin{array}{r} 14 \\ -\ 8 \\ \hline \end{array}$	32. $\begin{array}{r} 11 \\ -\ 3 \\ \hline \end{array}$	33. $\begin{array}{r} 11 \\ -\ 5 \\ \hline \end{array}$	34. $\begin{array}{r} 14 \\ -\ 9 \\ \hline \end{array}$	35. $\begin{array}{r} 13 \\ -\ 7 \\ \hline \end{array}$
36. $\begin{array}{r} 13 \\ -\ 6 \\ \hline \end{array}$	37. $\begin{array}{r} 17 \\ -\ 9 \\ \hline \end{array}$	38. $\begin{array}{r} 12 \\ -\ 5 \\ \hline \end{array}$	39. $\begin{array}{r} 12 \\ -\ 3 \\ \hline \end{array}$	40. $\begin{array}{r} 12 \\ -\ 6 \\ \hline \end{array}$
41. $\begin{array}{r} 13 \\ -\ 4 \\ \hline \end{array}$	42. $\begin{array}{r} 15 \\ -\ 7 \\ \hline \end{array}$	43. $\begin{array}{r} 10 \\ -\ 1 \\ \hline \end{array}$	44. $\begin{array}{r} 15 \\ -\ 8 \\ \hline \end{array}$	45. $\begin{array}{r} 13 \\ -\ 9 \\ \hline \end{array}$

3.8 Exchanging In Subtraction

The 45 examples you just completed along with the 55 examples in Practice Exercise 28 comprise the basic 100 combinations you need to know to do all subtraction examples. Be sure you know them well before continuing.

Example

The high temperature for today was 85° and the low was 66°. What is the difference in the two temperatures?

Solution

Subtract: 85° − 66° = ?

Problem	Step 1	Step 2	Step 3	Check
85 −66	7 1 ~~8~~5 − 6 6	7 1 ~~8~~5 − 6 6 9	7 1 ~~8~~5 − 6 6 1 9	66 +19 85
	You can't subtract 6 from 5, so *exchange* 1 ten in the minuend for 10 ones. Instead of 8 tens and 5 ones you now have 7 tens and 15 ones.	Subtract: 15 − 6 = 9	Subtract: 7 − 6 = 1	

Thus: 85 − 66 = 19.

Practice Exercise 36

Subtract. Follow the method just illustrated. The first two examples have been partially completed for you.

1. 81 ~~9~~6 −38
2. 31 ~~4~~2 −25
3. 62 −39
4. 45 −29
5. 64 −48
6. 37 −19
7. 82 −57
8. 51 −16
9. 64 −29
10. 56 −17

Example

Mary deposited $923 in the bank and wrote checks for $758. How much money has she left in her checking account?

Solution

To find out what remains in her checking account we must subtract \$758 from \$923.
Subtract: \$923 − 758 = ?

Problem	Step 1	Step 2	Step 3	Check
\$923 − 758	1 1 \$ 9 2 3 − 7 5 8 5	8 11 1 \$ 9 2 3 − 7 5 8 6 5	8 11 1 \$ 9 2 3 − 7 5 8 \$ 1 6 5	\$758 + 165 \$923
	Exchange 1 ten in the minuend for 10 ones since you can't subtract 8 from 3. You now have 1 ten and 13 ones instead of 2 tens and 3 ones. Subtract: 13 − 8 = 5	In the *tens* column you can't subtract 5 from 1, so exchange 1 hundred in the minuend for 10 tens. You now have 8 hundreds and 11 tens rather than 9 hundreds and 1 ten. Subtract: 11 − 5 = 6	Subtract: 8 − 7 = 1	

Thus: \$923 − \$758 = \$165.

Once you have mastered the idea of *exchanging* you can do all subtraction examples having any number of places. Look at the following examples before continuing to the practice exercise.

Problem	Step 1	Step 2	Step 3	Step 4	Check
3,814 −1,859	0 1 3 8 ~~1~~ 4 − 1 8 5 9 5	7 10 1 3 ~~8~~ ~~1~~ 4 − 1 8 5 9 5 5	2 17 10 1 ~~3~~ ~~8~~ ~~1~~ 4 − 1 8 5 9 9 5 5	2 17 10 1 ~~3~~ ~~8~~ ~~1~~ 4 − 1 8 5 9 1 9 5 5	1,859 +1,955 3,814
	In the *ones* column, you can't subtract 9 from 4, so *exchange* 1 ten in the minuend for 10 ones. You now have 0 tens and 14 ones rather than 1 ten and 4 ones. Subtract: $14 - 9 = 5$	In the *tens* column, you can't subtract 5 from 0, so *exchange* 1 hundred in the minuend for 10 tens. You now have 7 hundreds and 10 tens rather than 8 hundreds and 0 tens. Subtract: $10 - 5 = 5$	You can't subtract 8, from 7, so exchange 1 thousand in the minuend for 10 hundreds. You now have 2 thousands and 17 hundreds. Subtract: $17 - 8 = 9$	Subtract: $2 - 1 = 1$	

Thus: $3{,}814 - 1{,}859 = 1{,}955$.

Problem	Step 1	Step 2	Step 3	Check
704 −398	6 9 1 ~~7~~ ~~0~~ 4 − 3 9 8 6	6 9 1 ~~7~~ ~~0~~ 4 − 3 9 8 0 6	6 9 1 ~~7~~ ~~0~~ 4 − 3 9 8 3 0 6	398 +306 704
	In the *ones* column, you can't subtract 8 from 4, so exchange 1 ten in the minuend for 10 ones. Instead of 70 tens you now have 69 tens and 14 ones. Subtract: 14 − 8 = 6	In the *tens* column, subtract: 9 − 9 = 0	Subtract: 6 − 3 = 3	

Problem	Step 1	Step 2	Step 3	Check
400 −259	3 9 1 ~~4~~ ~~0~~ 0 − 2 5 9 1	3 9 1 ~~4~~ ~~0~~ 0 − 2 5 9 4 1	3 9 1 ~~4~~ ~~0~~ 0 − 2 5 9 1 4 1	259 +141 400
	You can't subtract 9 from 0, so exchange 1 ten in the minuend for 10 ones. Instead of 40 tens you now have 39 tens and 10 ones. Subtract: 10 − 9 = 1	Subtract: 9 − 5 = 4	Subtract: 3 − 2 = 1	

Practice Exercise 37

In each of the following examples illustrate the method of exchanging in subtraction.

1. 6 tens and 3 ones = 5 tens and ________ ones
2. 2 tens and 0 ones = 1 ten and ________ ones
3. 8 tens and 4 ones = 7 tens and ________ ones
4. 3 hundreds and 4 tens = 2 hundreds and ________ tens
5. 6 hundreds and 2 tens = 5 hundreds and ________ tens

Find the difference or remainders in each of the following problems. Check by adding.

6.	7.	8.	9.	10.
383	996	535	684	209
− 56	− 48	−286	−455	− 98

11.	12.	13.	14.	15.
200	100	1,690	6,975	1,023
− 98	− 37	− 956	−3,821	− 394

Write each of the following problems in column form before subtracting.

16. 5,303 − 4,729 =
17. 5,970 − 5,796 =
18. 9,895 − 5,599 =
19. 7,290 − 329 =
20. 915 − 92 =

Term You Should Remember

Exchange To give in return for an equivalent quantity.

Review Of Important Ideas

Some important ideas covered in Chapter 3 so far are

Subtraction is "taking away."

The symbol − means to subtract.

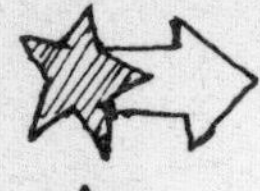
Digits having the same place values are lined up in columns before subtracting.

Numbers are subtracted by pairs only.

We *exchange* when it is impossible to subtract in any column.

3.9 More Difficult Word Problems Requiring Subtraction

Before we attempt to solve any more problems in mathematics, let's look again at some special hints that may help us become better problem solvers.

1. Develop habits of careful reading in math. These habits result from reading critically (with questioning mind) and from practice. This book gives much practice in reading directions, explanations, and problems. Naturally a GED test will not have these explanations, but your practice will help you to figure out problems on your own.
2. Do not be alarmed if your reading rate in mathematics is slow — it should be. Often you must reread to be sure you understand. Directions should be read slowly so that you know what is expected. Many people skip over directions quickly because they are eager to start on the problems. But what a great waste it is when work is done incorrectly because the directions were ignored or read carelessly. Directions often include sample problems to show how the work is done and how the answer is to be chosen.
3. Read mathematics with thought and questions. As you read, ask yourself:
    1. What is given? (facts in problem)
    2. What is unknown? (question to be answered)
    3. How should this be solved? (method to be used to find the answer)

This is a much different style of reading from what you might use when you look through a newspaper or magazine. In mathematics, directions, explanations, and problems are compressed into very few words. Each word is therefore very important and cannot be overlooked by speedy reading.

Try to keep some of these suggestions in mind when you approach this problem and all others that follow.

Example

Mary earned $135 last week for 5 days of work. This week her paycheck was $162 since she worked some overtime hours. How much more did she earn this week than last week?

Solution

1. What is given?
   You are told Mary's two earnings, $135 for last week and $162 this week.
2. What is unknown?
   You are asked to find out how much more she earned this week than last week.
3. How should this be solved?
   The problem asks for the *difference* in the two paychecks, so subtract $135 from $162.

$$\begin{array}{r} \$162 \\ -\ 135 \\ \hline \$\ 27 \end{array}$$

Mary earned $27 more this week than last week.

Practice Exercise 38

1. You buy a wallet for $8 and give the cashier a $20 bill. How much change do you get?

2. A man weighed 218 lb. before he dieted. If he lost 39 lb., how much does he now weigh?

3. A lawn mower usually sells for $498. Its sale price is $449. How much would you save buying it during the sale?

4. The local paper announced that 7,500 lb. of haddock were caught this week and 3,600 lb. last week. How many more pounds were caught this week?

5. Mr. Smith buys a new house for $25,125 and sells his old house for $17,750. How much more does his new house cost?

6. A cashier is given $25 at the start of each day to help her make change. At the end of the day, she counts her money and finds that it totals $804. How much money did she collect during the entire day?

7. Two sides of a triangle are 27 ft. and 32 ft. Find the length of the third side if the perimeter is 100 ft. (*Hint:* First add the two known sides together before subtracting the sum from the perimeter.)

8. The distance from Omaha to Tulsa to Houston is 510 mi. If the distance from Omaha to Tulsa is 385 mi., how far is it from Tulsa to Houston?

9. An automobile company produced 8,634 cars in February. How many cars did they produce in March if the monthly output was 796 cars less than in February?

10. A dance was held in a social hall on two successive Saturday nights. If the attendance on the first night was 240 couples and on the second night was 229 couples, how many more couples appeared on the first night?

Let's See How Your Skill In Subtraction Has Improved

This chapter proceeded from simple to more complex subtraction examples. You have probably found your past skills returning and understand the problems better than you did before. The posttest will help check your progress.

Posttest 3

Write your answer in the spaces provided.

1.	2.	3.	4.	5.
9 − 2	94 − 73	563 − 361	478 − 120	78 − 5

6.	7.	8.	9.	10.
365 − 43	13 − 7	65 − 9	436 − 282	743 − 485

11. 8,076 − 2,341

12. 3,004 − 2,877 =

13. 5,000 − 309

14. What is the difference between 63¢ and 96¢?

15. If a dress manufacturer ships 2,390 dresses as a partial order of 3,265 dresses, how many more must he ship?

16. How much change do you receive from a 20-dollar bill if your purchases total 13 dollars 56 cents?

17. From 1,812 subtract 345.

18. If 101 ft. of wire is needed to enclose a triangular plot of ground, find the length of the third side if the sum of the other two sides is 66 ft.

19. This house has been reduced in price to $19,995. How much money can be saved buying the house at the new price?

20. Show the result of subtracting 3 from 7 on the number line below.

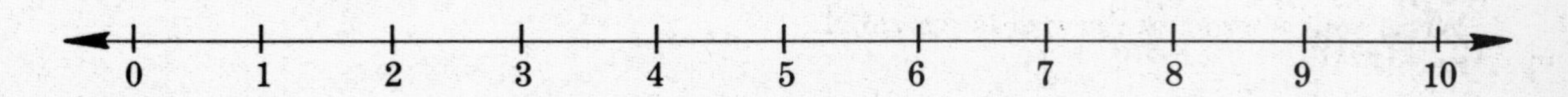

ANSWERS AND EXPLANATIONS TO POSTTEST 3

1. 7	**2.** 21	**3.** 202	**4.** 358	**5.** 73
6. 322	**7.** 6	**8.** 56	**9.** 154	**10.** 258
11. 5,735	**12.** 127	**13.** 4,691	**14.** 96¢ − 63¢ = 33¢	**15.** 3,265 − 2,390 = 875

16. 6 dollars and 44 cents

17.
```
  1,812
-   345
  1,467
```

18.
```
  101 ft.
-  66 ft.
   35 ft.
```

19.
```
 $24,000
 -19,995
 $ 4,005
```

20.

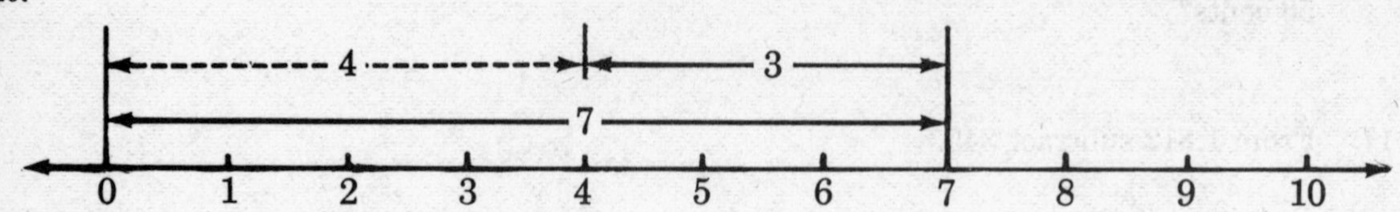

A Score of	*Means That You*
18–20	Did very well. Move on to Chapter 4.
16–17	Know the material except for a few points. Reread the section about the ones you missed.
13–15	Need to check carefully the sections you missed.
0–12	Need to review the chapter again to refresh your memory and improve your skills.

Questions	*Are Covered in Section*
1–4	3.2
20	3.3
5, 6	3.4
14	3.5
16	3.6
7	3.7
8–13	3.8
15, 17–19	3.9

Hold It!

Now that you have reviewed all the combinations in subtraction, go back to Practice Exercises 28 and 35 and see how well you can now do on the 100 combinations. Can you make a perfect score?

Puzzle Time

Did you enjoy the other cross-number puzzle? Here is one which will review *subtraction of whole numbers*. The solution is on page 93.

ACROSS

1. 469 − 234 =
2. 78 − 19 =
4. 532 − 278 =
6. 481 − 206 =
8. 4,610 − 2,845 =
10. 300 − 162 =
12. 82 − 64 =
13. 106 − 89 =
16. 704 − 589 =
18. 2,811 − 1,993 =
22. 201 − 185 =
23. 100 − 48 =
26. 482 − 293 =
27. 23,051 − 15,838 =
29. 2,000 − 1,367 =
30. 2,005 − 1,755 =
31. 83 − 6 =
32. 602 − 197 =

DOWN

1. 790 − 589 =
3. 1,071 − 143 =
4. 8,970 − 6,390 =
5. 6,460 − 2,347 =
7. 100 − 48 =
9. 250 − 189 =
11. 500 − 463 =
14. 63 + 37 − 9 =
15. 57 − 19 =
17. 81 − 23 =
19. 71 − 57 =
20. 4,040 − 2,348 =
21. 11,068 − 2,498 =
22. 97 − 79 =
24. 100 − 78 =
25. 71 − 35 =
26. 340 − 203 =
28. 1,827 − 1,432 =

ANSWERS FOR CHAPTER 3

PRETEST 3

1. 3	**2.** 22	**3.** 222	**4.** 201	**5.** 216
6. 93	**7.** 6	**8.** 65	**9.** 154	**10.** 245
11. 915	**12.** 18	**13.** 3,792	**14.** 11¢	**15.** $36,489
16. 5 dollars 18 cents		**17.** 420	**18.** $18,195	**19.** 59 ft.

20.

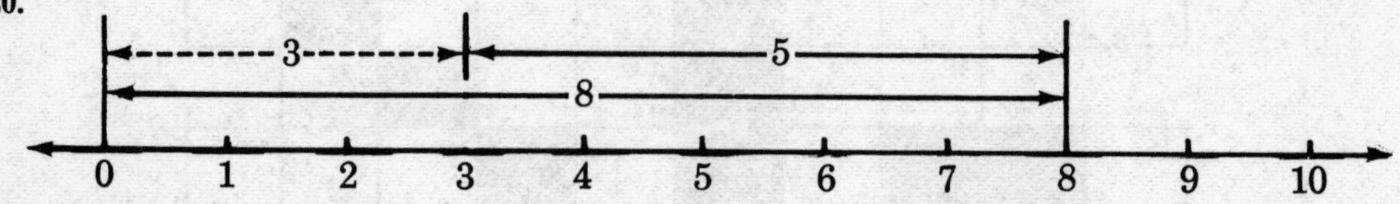

PRACTICE EXERCISE 28

1. 5	**2.** 0	**3.** 5	**4.** 6	**5.** 5
6. 0	**7.** 2	**8.** 1	**9.** 1	**10.** 0
11. 1	**12.** 7	**13.** 1	**14.** 3	**15.** 1
16. 6	**17.** 0	**18.** 6	**19.** 1	**20.** 0
21. 1	**22.** 2	**23.** 4	**24.** 2	**25.** 7
26. 2	**27.** 3	**28.** 2	**29.** 2	**30.** 0
31. 1	**32.** 5	**33.** 0	**34.** 3	**35.** 3
36. 4	**37.** 8	**38.** 3	**39.** 5	**40.** 3
41. 6	**42.** 2	**43.** 3	**44.** 7	**45.** 0
46. 9	**47.** 4	**48.** 8	**49.** 0	**50.** 4
51. 2	**52.** 4	**53.** 1	**54.** 4	**55.** 0

PRACTICE EXERCISE 29

1. 12	**2.** 35	**3.** 11	**4.** 32	**5.** 63
6. 73	**7.** 23	**8.** 57	**9.** 41	**10.** 25

PRACTICE EXERCISE 30

1. 25	**2.** 325	**3.** 1,602	**4.** 2,110	**5.** 411,080
6. 14	**7.** 202	**8.** 397	**9.** 22,235	**10.** $1,233,000
11. 225	**12.** 6,420	**13.** 227	**14.** 245	**15.** 232,055

PRACTICE EXERCISE 31

1.

6 3
9
0 1 2 3 4 5 6 7 8 9 10

2.

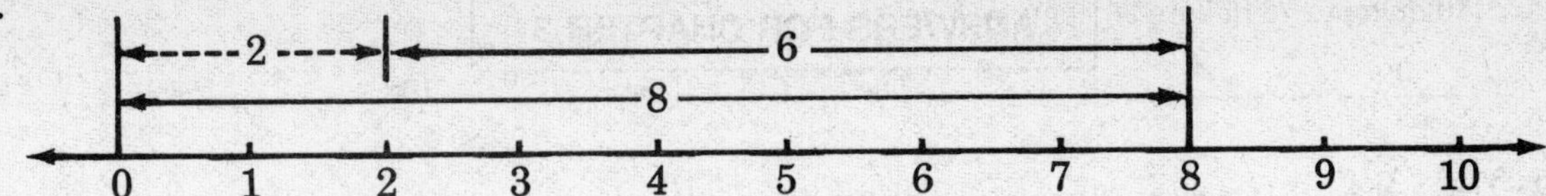

3.

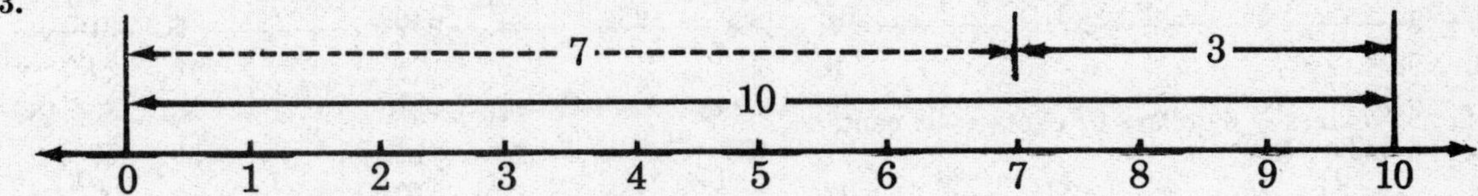

4.

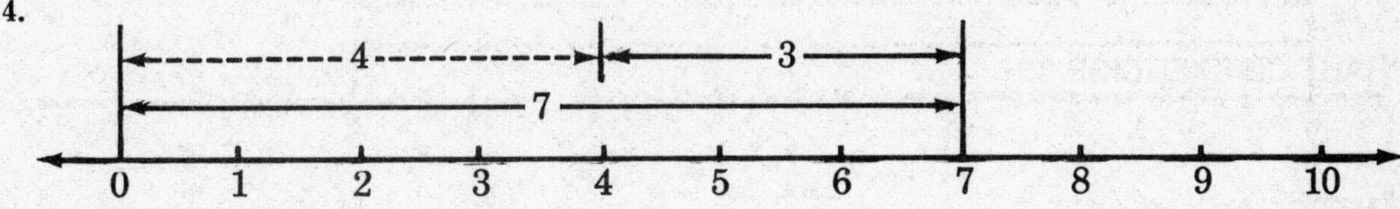

5.

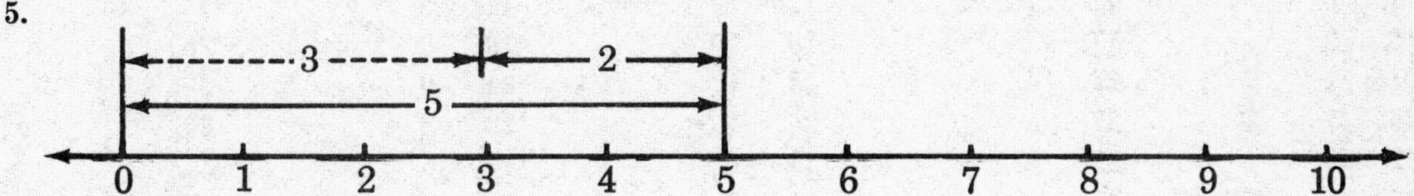

PRACTICE EXERCISE 32

1.	31	**2.**	327	**3.**	514	**4.**	4,212	**5.**	6,704
6.	471	**7.**	5,412	**8.**	375	**9.**	45	**10.**	$420,000

PRACTICE EXERCISE 33

1.	$8	− 5	$3
2.	$73	− 51	$22
3.	$29	− 21	$ 8
4.	338,400 lb.	−231,200 lb.	107,200 lb.
5.	$8,764	− 6,560	$2,204
6.	$29,900	− 26,000	$ 3,900
7.	45	−31	14
8.	$1,195	− 1,075	$ 120
9.	$25	− 14	$11
10.	43	−22	21

PRACTICE EXERCISE 34

1. 5 dollars − 3 dollars 74 cents = 1 dollar 26 cents

___0___ $5; ___1___ $1; ___0___ 50¢; ___1___ 25¢; ___0___ 10¢; ___0___ 5¢; ___1___ 1¢

2. 2 dollars − 1 dollar 59 cents = 41 cents

___0___ $5; ___0___ $1; ___0___ 50¢; ___1___ 25¢; ___1___ 10¢; ___1___ 5¢; ___1___ 1¢

3. 10 dollars − 9 dollars 95 cents = 5 cents

0 $5; 0 $1; 0 50¢; 0 25¢; 0 10¢; 1 5¢; 0 1¢

4. 20 dollars − 11 dollars 43 cents = 8 dollars 57 cents

1 $5; 3 $1; 1 50¢; 0 25¢; 0 10¢; 1 5¢; 2 1¢

5. 7 dollars − 6 dollars 67 cents = 33 cents

0 $5; 0 $1; 0 50¢; 1 25¢; 0 10¢; 1 5¢; 3 1¢

PRACTICE EXERCISE 35

1.	3	**2.**	9	**3.**	4	**4.**	8	**5.**	9
6.	5	**7.**	5	**8.**	8	**9.**	9	**10.**	3
11.	6	**12.**	7	**13.**	5	**14.**	8	**15.**	4
16.	9	**17.**	7	**18.**	2	**19.**	3	**20.**	8
21.	1	**22.**	4	**23.**	6	**24.**	9	**25.**	7
26.	7	**27.**	5	**28.**	9	**29.**	8	**30.**	2
31.	6	**32.**	8	**33.**	6	**34.**	5	**35.**	6
36.	7	**37.**	8	**38.**	7	**39.**	9	**40.**	6
41.	9	**42.**	8	**43.**	9	**44.**	7	**45.**	4

PRACTICE EXERCISE 36

1.	58	**2.**	17	**3.**	23	**4.**	16	**5.**	16
6.	18	**7.**	25	**8.**	35	**9.**	35	**10.**	39

PRACTICE EXERCISE 37

1.	13 ones	**2.**	10 ones	**3.**	14 ones	**4.**	14 tens	**5.**	12 tens
6.	327	**7.**	948	**8.**	249	**9.**	229	**10.**	111
11.	102	**12.**	63	**13.**	734	**14.**	3,154	**15.**	629
16.	574	**17.**	174	**18.**	4,296	**19.**	6,961	**20.**	823

PRACTICE EXERCISE 38

1. $20 − 8 = $12

2. 218 lb. − 39 lb. = 179 lb.

3. $498 − 449 = $ 49

4. 7,500 lb. − 3,600 lb. = 3,900 lb.

5. $25,125 − 17,750 = $ 7,375

6. $804 − 25 = $779

7. 27 ft. + 32 ft. = 59 ft.; 100 ft. − 59 ft. = 41 ft.

8. 510 mi. − 385 mi. = 125 mi.

9. 8,634 − 796 = 7,838

10. 240 − 229 = 11

CROSS-NUMBER PUZZLE SOLUTION

(1) 2	3	5				(2) 5	(3) 9		
0	■	■	(4) 2	5	(5) 4	■	(6) 2	7	(7) 5
(8) 1	7	(9) 6	5	■	(10) 1	(11) 3	8	■	2
		(12) 1	8	■	(13) 1	7	■	(14) 9	
(15) 3	■	■	0	■	3	■	(16) 1	1	(17) 5
(18) 8	(19) 1	8	■	(20) 1	■	(21) 8	■	■	8
	4	■	(22) 1	6	■	(23) 5	(24) 2		
(25) 3	■	(26) 1	8	9	■	(27) 7	2	1	(28) 3
(29) 6	3	3	■	(30) 2	5	0	■	■	9
		(31) 7	7				(32) 4	0	5

Another basic operation in mathematics is multiplication. We use multiplication when we want to find the price of more than one item. The cost of 2 qts. of milk at 41c a quart is two times 41¢. The cost of 3 cans of juice at 43¢ a can is three times 43¢.

At first glance you would say it looks like addition. You are right. Multiplication is really repeated addition, but by a faster method. For example, 2 times 41 means 41 + 41. In an example such as 2 times 41, adding and multiplying are equally fast. In an example such as 8 times 28, addition is a slow process compared to multiplication. The larger the numbers, the more useful multiplication becomes.

See What You Know And Remember — Pretest 4

This test gives you a chance to see how good you are in multiplication. Do all your work in the spaces provided.

1.	2.	3.	4.	5.
$\begin{array}{r} 7 \\ \times 5 \\ \hline \end{array}$	$\begin{array}{r} 34 \\ \times\ 2 \\ \hline \end{array}$	$\begin{array}{r} 212 \\ \times\ \ 0 \\ \hline \end{array}$	$\begin{array}{r} 73 \\ \times\ 3 \\ \hline \end{array}$	$\begin{array}{r} 402 \\ \times\ \ 6 \\ \hline \end{array}$

6. $\begin{array}{r} 330 \\ \times\ 2 \\ \hline \end{array}$

7. $\begin{array}{r} 35 \\ \times 12 \\ \hline \end{array}$

8. $\begin{array}{r} 43 \\ \times 34 \\ \hline \end{array}$

9. $\begin{array}{r} 56 \\ \times 47 \\ \hline \end{array}$

10. $\begin{array}{r} 306 \\ \times\ 93 \\ \hline \end{array}$

11. $\begin{array}{r} 510 \\ \times\ 70 \\ \hline \end{array}$

12. $\begin{array}{r} 703 \\ \times\ 30 \\ \hline \end{array}$

13. $\begin{array}{r} 500 \\ \times 200 \\ \hline \end{array}$

14. $\begin{array}{r} 378 \\ \times\ 98 \\ \hline \end{array}$

15. How many cents do 6 cans of juice at 43¢ a can cost?

16. Find the product of 3 × 3 on the number line.

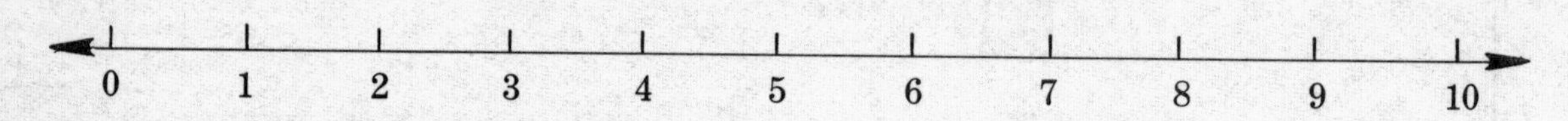

17. A weekly newspaper costs 25¢. If you purchase the paper each week for 50 weeks, what will be your total cost in cents?

18. What is the cost (in cents) of 4 lbs. of chicken breast and 2 lbs. of chicken legs?

THE BREWSTER MEAT SHOP	
Chicken breasts	85¢ lb.
Chicken legs	65¢ lb.

19. A car travels 9 mi. for each gallon of gasoline. If you purchase 16 gal. of gasoline, how many miles can you travel?

20. A dress manufacturer makes and sells dresses for $32 each. If he packs one to a box and ships 1,032 boxes, how much money will he receive?

21. Martin is covering his rectangular kitchen floor with linoleum. How many square feet of material must he purchase if his room measures 6 ft. by 13 ft.?

22. Using the formula for the perimeter of a rectangle, $P = 2L + 2W$, find P if $L = 14$ in. and $W = 8$ in.

Check your answers by turning to the end of the chapter. Add up all that you had correct.

A Score of	*Means That You*
19–22	Did very well. You can move to Chapter 5.
17–18	Know this material except for a few points. Read the sections about the ones you missed.
14–16	Need to check carefully on the sections you missed.
0–13	Need to work with this chapter to refresh your memory and improve your skills.

Questions	*Are Found in Section*
16	4.2
1	4.4
2–6	4.5
15, 18, 19	4.6
7–14	4.7
21, 22	4.8
17, 20	4.9

4.1 Multiplication Is Repeated Addition

You may be wondering why we said that *multiplication is repeated addition.* Let's look into the meaning of this statement.

When you add the same term a given number of times — for example, "Add 6 three times" — the problem is written as

$$\begin{array}{r} 6 \\ 6 \\ +6 \\ \hline 18 \end{array}$$

Since 6 is an addend *three* times, you can write it as a *multiplication* example:

$$\begin{array}{r} 6 \\ \times 3 \\ \hline 18 \end{array}$$

This is pictured as 3 groups of 6 items:

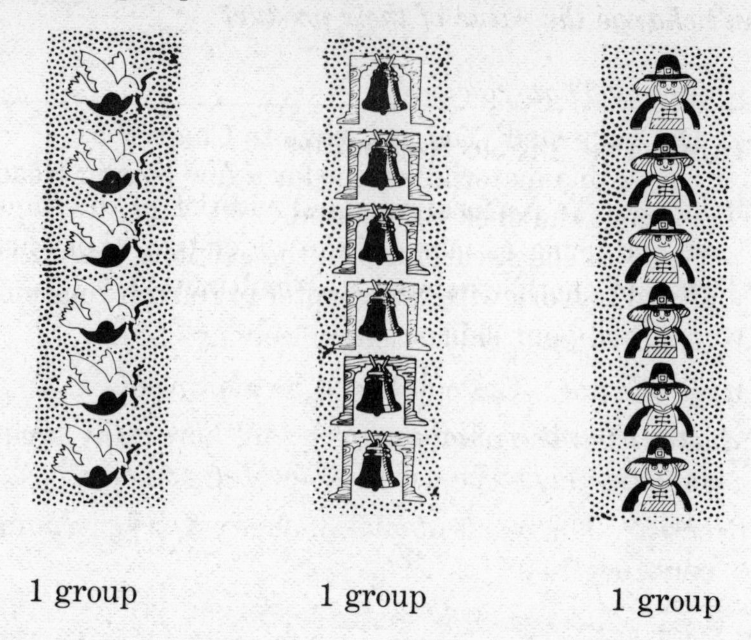

1 group 1 group 1 group

Multiplication is repeated addition but much faster than adding the same number many times. The symbol × is called the *times* sign, so whenever you see that sign you know to perform the operation of multiplication.

There is a language for multiplication too. Look at this example:

The *times* sign tells us to multiply ⟶

$$\begin{array}{r l} 6 & \leftarrow \text{multiplicand} \\ \times 3 & \leftarrow \text{multiplier} \\ \hline 18 & \leftarrow \text{product} \end{array}$$

If you were to say, "Add 3 six times," you would write it as $3 + 3 + 3 + 3 + 3 + 3$. Since 3 is the addend six times, you can write it as a multiplication example:

$$\begin{array}{r l} 3 & \leftarrow \text{multiplicand} \\ \times 6 & \leftarrow \text{multiplier} \\ \hline 18 & \leftarrow \text{product} \end{array}$$

This is pictured as 6 groups of 3 items.

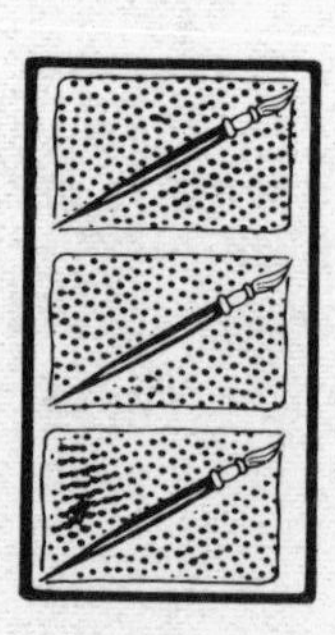

Therefore, $3 \times 6 = 6 \times 3$, since each product is 18. *Changing the order of the numbers doesn't change the value of their product.*

Terms You Should Remember

Multiply To perform repeated addition of the same addend using as many of the addends as there are units in another number (the multiplier).

Multiplication The act of multiplying.

Multiplicand The number to be multiplied.

Multiplier The number which tells how many times an addend is to be used in repeated addition.

Product The result of multiplying two or more numbers together.

Times The sign × of multiplication.

4.2 Help From The Number Line

One number can be added to another and one number can be subtracted from another. This joining of two things is called a *binary* operation. Since you also multiply two numbers together at a time, *multiplication* is a binary operation.

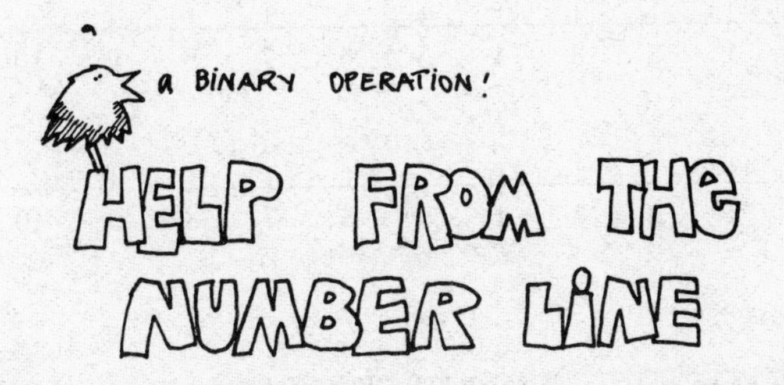

Multiplication can be shown on the number line. To multiply 3×2, represent this as a repeated addition of $2 + 2 + 2$ on the number line. Beginning at 0, draw a line segment 2 units to the right. From that point continue with another line segment also 2 units in length. This now brings you to a value corresponding to the point 4. Complete the addition of the three terms at the point which is 6 units to the right of 0. Thus the product of $3 \times 2 = 6$.

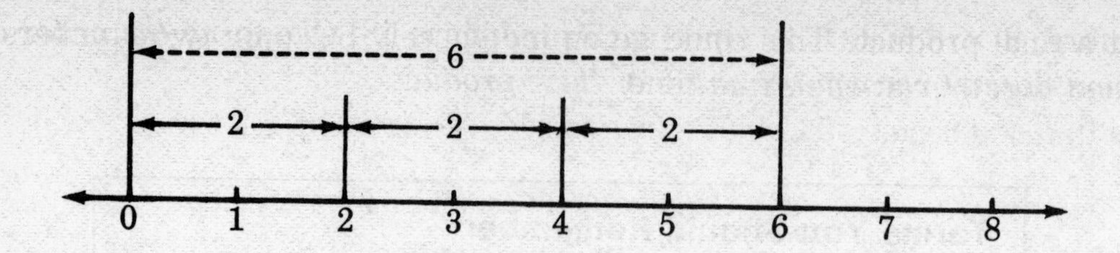

Multiplying 2×3 is pictured on the number line below. It is the repeated addition of $3 + 3$. As you see, the final result is 6, the same as 3×2. This further illustrates that changing the order of the numbers doesn't change the value of the product.

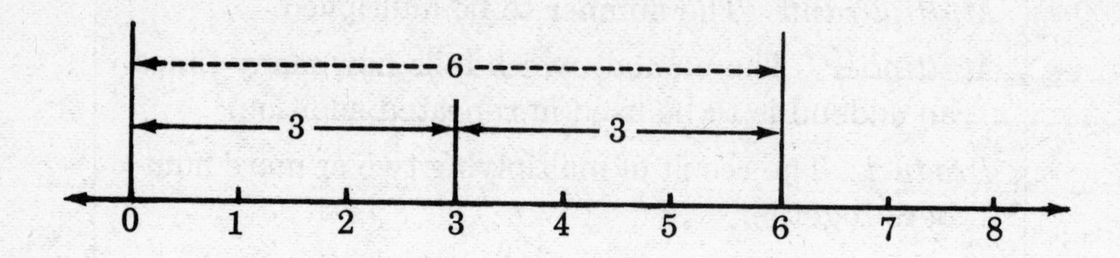

Practice Exercise 39

Illustrate each of the following multiplication problems on a number line and find the product.

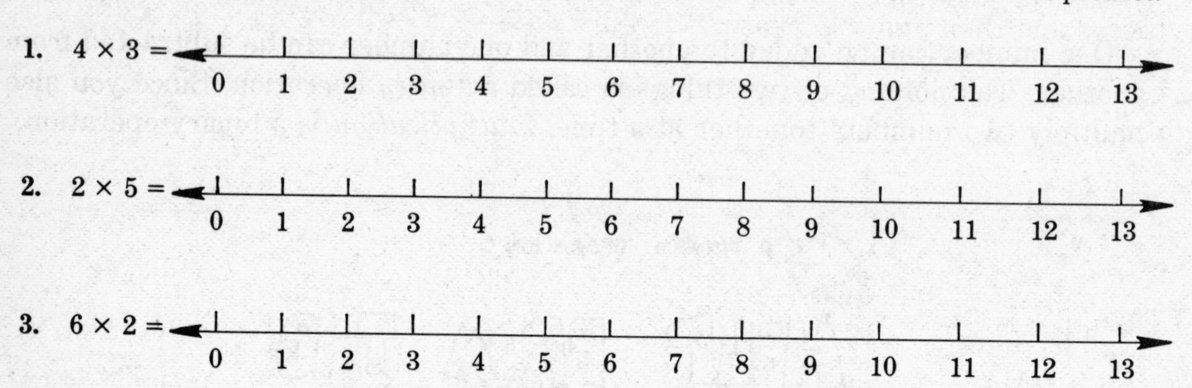

Does this mean that you can't multiply more than two numbers together? No, it doesn't mean that. When you added more than two numbers, you added two together and then added another to that sum. You do the same thing when multiplying.

For example, what is the product of $2 \times (3 \times 4)$?

First, find the product of the 3×4 inside the parentheses and then multiply that product by 2. Keep in mind that multiplication is a *binary operation* — an operation performed on two numbers at a time. For instance, in the above example multiply two of the numbers first, get a product, multiply that product by the next number and get a final product. This could go on indefinitely but only two numbers are multiplied together at the same time.

Let's find the result of $2 \times (3 \times 4)$, using the number line. First, 3×4:

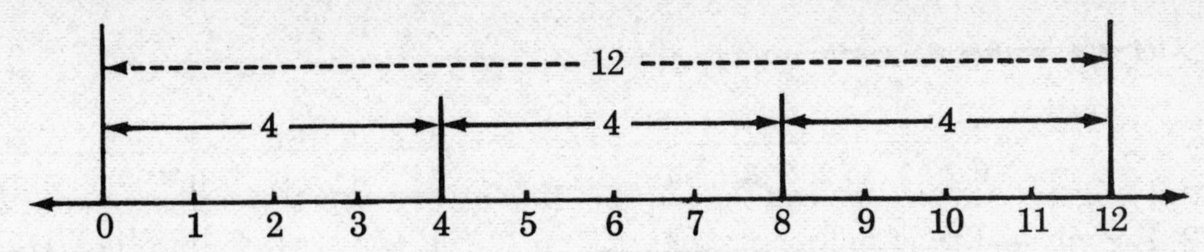

Then, 2×12:

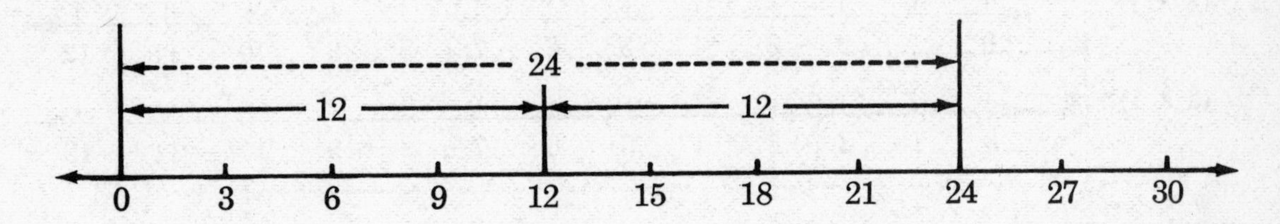

Thus you see that $3 \times 4 = 12$ and $2 \times 12 = 24$. Therefore,

$$2 \times (3 \times 4) = 24$$

Will the order of the multiplications matter? Suppose you were seeking an answer to $(2 \times 3) \times 4$. Here you start with the product of 2×3 inside the parentheses and then multiply that product by 4. First, 2×3:

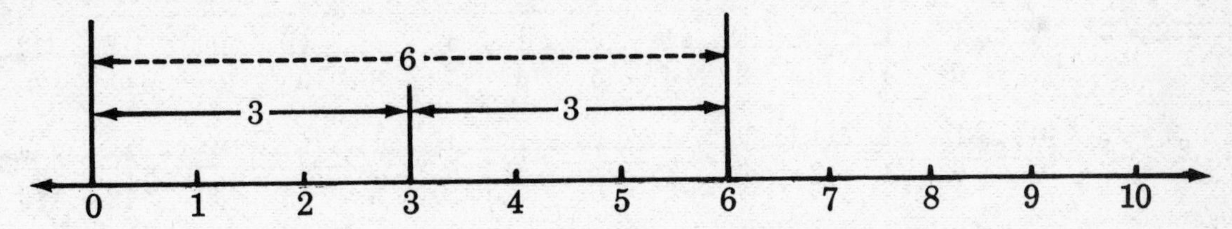

Then, 6×4:

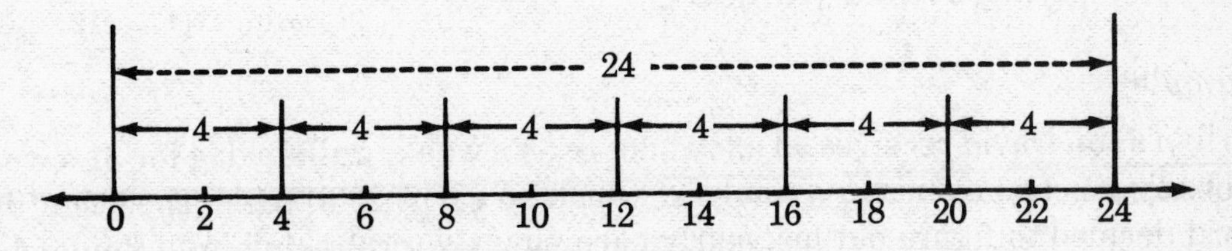

Thus you see that $2 \times 3 = 6$ and $6 \times 4 = 24$. Therefore,

$$(2 \times 3) \times 4 = 24$$

If three numbers are multiplied together, the product will be the same, regardless of the grouping order. That is,

$$(2 \times 3) \times 4 = 2 \times (3 \times 4)$$

Practice Exercise 40

Illustrate the truth of each of the following problems on the number line and find the product.

1. $3 \times (1 \times 4) = (3 \times 1) \times 4$

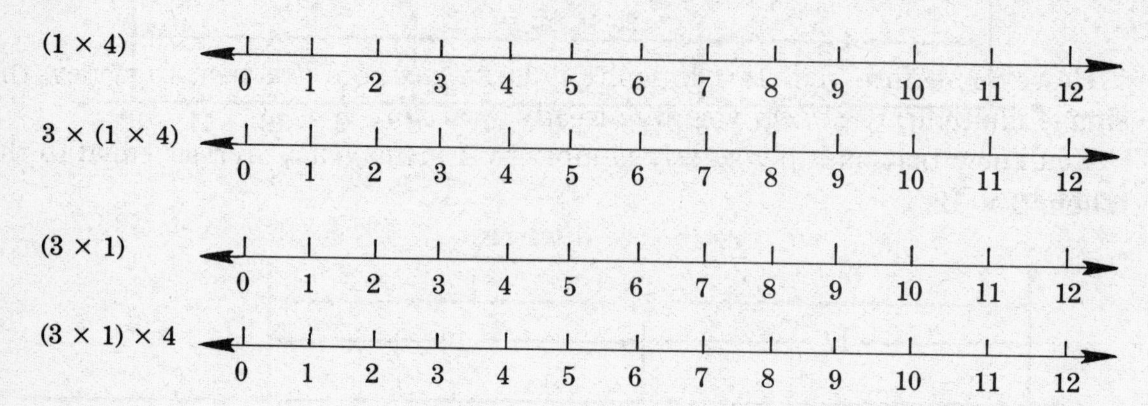

2. $(2 \times 2) \times 3 = 2 \times (2 \times 3)$

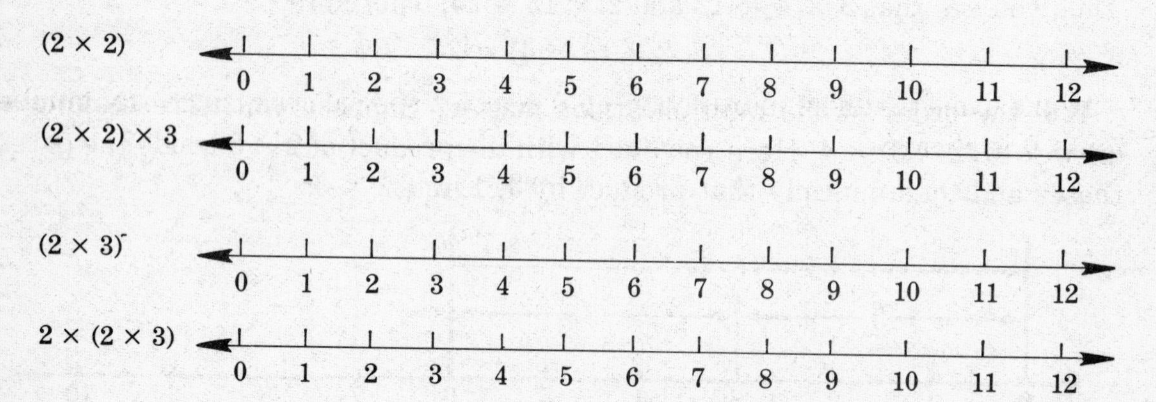

4.3 Multiplying Whole Numbers

Example

Marilyn's son David receives an allowance of \$4 a week, walks a dog for \$1 a week, and baby-sits two evenings a week for which he earns \$5 a week. The other day, David decided to figure out his yearly earnings. Can you help him? (Remember, there are 52 weeks in a year.)

Solution

David knew that he could find the answer in a simple way. He could write $4 + 4 + 4 \ldots$ in column form 52 times, and $1 + 1 + 1 \ldots$ 52 times and finally $5 + 5 + 5 \ldots$ 52 times. But he knew that repeated additions, such as adding \$4 a total of 52 times, can be more easily done by multiplication. He set up these three examples:

$$\$4 \times 52 = ?$$
$$\$1 \times 52 = ?$$
$$\$5 \times 52 = ?$$

However, before you start to multiply large numbers it is best to review the simple multiplication facts you are already aware of.

You know that multiplying any number by 1 gives you a product equal to the number.

$$\begin{array}{r} 1 \\ \times 3 \\ \hline \end{array} = \overbrace{1 + 1 + 1}^{\text{3 terms}} = 3$$

or

$$\begin{array}{r} 4 \\ \times 1 \\ \hline \end{array} = 4 = \overbrace{4}^{\text{1 term}} = 4$$

In the same manner, multiplying by 0 is also a special case, since any number times 0 is always 0.

$$\begin{array}{r} 0 \\ \times 6 \\ \hline \end{array} = \overbrace{0 + 0 + 0 + 0 + 0 + 0}^{\text{6 terms}} = 0$$

or

$$\begin{array}{r} 4 \\ \times 0 \\ \hline \end{array} = \overbrace{0}^{\text{0 terms}} = 0$$

This practice exercise contains only the special multipliers of 1 and 0. Even though these examples are simple, do them carefully.

Practice Exercise 41

Place the product under each example. Try for a perfect score.

1.	2.	3.	4.	5.	6.
$\begin{array}{r} 1 \\ \times 0 \\ \hline \end{array}$	$\begin{array}{r} 0 \\ \times 7 \\ \hline \end{array}$	$\begin{array}{r} 6 \\ \times 0 \\ \hline \end{array}$	$\begin{array}{r} 0 \\ \times 2 \\ \hline \end{array}$	$\begin{array}{r} 5 \\ \times 0 \\ \hline \end{array}$	$\begin{array}{r} 0 \\ \times 6 \\ \hline \end{array}$

7. $\begin{array}{r}0\\ \times 3\\ \hline\end{array}$	**8.** $\begin{array}{r}4\\ \times 0\\ \hline\end{array}$	**9.** $\begin{array}{r}0\\ \times 9\\ \hline\end{array}$	**10.** $\begin{array}{r}3\\ \times 0\\ \hline\end{array}$	**11.** $\begin{array}{r}2\\ \times 1\\ \hline\end{array}$	**12.** $\begin{array}{r}1\\ \times 1\\ \hline\end{array}$
13. $\begin{array}{r}0\\ \times 1\\ \hline\end{array}$	**14.** $\begin{array}{r}4\\ \times 1\\ \hline\end{array}$	**15.** $\begin{array}{r}9\\ \times 1\\ \hline\end{array}$	**16.** $\begin{array}{r}1\\ \times 5\\ \hline\end{array}$	**17.** $\begin{array}{r}8\\ \times 0\\ \hline\end{array}$	**18.** $\begin{array}{r}0\\ \times 8\\ \hline\end{array}$
19. $\begin{array}{r}0\\ \times 0\\ \hline\end{array}$	**20.** $\begin{array}{r}9\\ \times 0\\ \hline\end{array}$	**21.** $\begin{array}{r}0\\ \times 5\\ \hline\end{array}$	**22.** $\begin{array}{r}7\\ \times 0\\ \hline\end{array}$	**23.** $\begin{array}{r}3\\ \times 1\\ \hline\end{array}$	**24.** $\begin{array}{r}1\\ \times 8\\ \hline\end{array}$
25. $\begin{array}{r}1\\ \times 2\\ \hline\end{array}$	**26.** $\begin{array}{r}1\\ \times 9\\ \hline\end{array}$	**27.** $\begin{array}{r}0\\ \times 4\\ \hline\end{array}$	**28.** $\begin{array}{r}1\\ \times 3\\ \hline\end{array}$	**29.** $\begin{array}{r}5\\ \times 1\\ \hline\end{array}$	**30.** $\begin{array}{r}2\\ \times 0\\ \hline\end{array}$
31. $\begin{array}{r}1\\ \times 6\\ \hline\end{array}$	**32.** $\begin{array}{r}1\\ \times 4\\ \hline\end{array}$	**33.** $\begin{array}{r}8\\ \times 1\\ \hline\end{array}$	**34.** $\begin{array}{r}6\\ \times 1\\ \hline\end{array}$	**35.** $\begin{array}{r}1\\ \times 7\\ \hline\end{array}$	**36.** $\begin{array}{r}7\\ \times 1\\ \hline\end{array}$

4.4 Those Old Combinations Again

Skill and speed in multiplication depend very much on knowing how numbers combine. There are basic number combinations using all the symbols from 0 to 9. These are well worth practicing to gain skill because they save you so much work when you know them. They are sometimes called the multiplication tables.

There are 10 combinations for each number, or 100 altogether. Thirty-six of them were done by you in the last practice exercise. With the 64 more in this next exercise, you will have learned the hundred combinations in multiplication.

Once again we suggest that you do not write the answers in the book. Write your answers on a separate piece of paper so you can use this exercise for practice as many times as you wish. Don't write the problem. Place your paper under the first row and write each answer. Then fold back the paper and place it under the next row. Write the answers, fold back the paper, and so forth, until you have all eight rows. The idea is to see how you can improve by recognizing the answers more and more quickly. Keep your own record of this.

Practice Exercise 42

Find the product for each of the following problems. Remember not to write in the book, so that you can practice again and again.

1. 6 ×5
2. 2 ×9
3. 7 ×2
4. 5 ×9
5. 7 ×8
6. 6 ×4
7. 6 ×7
8. 6 ×9
9. 8 ×6
10. 7 ×7
11. 7 ×9
12. 9 ×7
13. 8 ×5
14. 4 ×4
15. 5 ×6
16. 3 ×7
17. 4 ×9
18. 5 ×4
19. 8 ×3
20. 7 ×6
21. 7 ×3
22. 4 ×7
23. 9 ×9
24. 8 ×8
25. 9 ×3
26. 5 ×7
27. 5 ×8
28. 4 ×6
29. 9 ×4
30. 3 ×9
31. 2 ×8
32. 4 ×8
33. 6 ×6
34. 2 ×5
35. 3 ×5
36. 8 ×7
37. 7 ×4
38. 9 ×6
39. 5 ×5
40. 9 ×2
41. 9 ×8
42. 5 ×3
43. 8 ×9
44. 2 ×7
45. 6 ×8
46. 2 ×2
47. 8 ×4
48. 3 ×2
49. 2 ×3
50. 9 ×5
51. 2 ×4
52. 3 ×3
53. 4 ×5
54. 5 ×2
55. 3 ×6
56. 3 ×8
57. 8 ×2
58. 7 ×5
59. 6 ×2
60. 3 ×4
61. 2 ×6
62. 4 ×3
63. 6 ×3
64. 4 ×2

4.5 Multiplying By A One-Digit Number

Example

At the year-end sale, Mrs. De Vogt bought 3 needlepoint designs for $12 each. How much did they cost her?

Needlepoint
& Crewel

Solution

The problem is simply the repeated addition of $12 + $12 + $12, or $36. You know that a problem of this type can be done more easily and more quickly by multiplication. How do you multiply $12 × 3?

The number 12 can be expressed as 10 + 2; and 3 × 12 can be written as 3 × (10 + 2). The product is 36. If you look at this problem from a different angle, you see that 3 × 10 = 30 and 3 × 2 = 6; the sum of these two products 30 + 6 = 36.

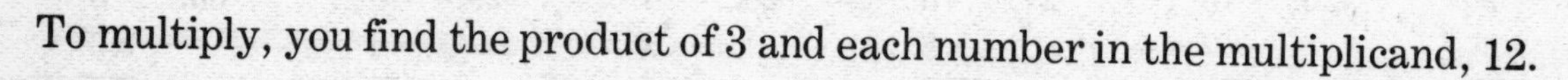

To multiply, you find the product of 3 and each number in the multiplicand, 12.

Problem	Step 1	Step 2	Step 3
12 × 3	1 2 × 3 6	1 2 × 3 6 3 0	1 2 × 3 6 + 3 0 3 6
	Multiply: 3 × 2 = 6 This is called a *partial* product.	Multiply: 3 × 1 (Recall that a 1 in the *tens* place is a 10.) This is the other *partial* product. 3 × 10 = 30	Add the partial products together: 30 + 6 = 36

Thus: 12 × 3 = 36.

Now that you know the *why* for multiplying numbers, let's look for a quicker method of multiplying.

Problem	Step 1	Step 2
$\begin{array}{r} 12 \\ \times\ 3 \\ \hline \end{array}$	$\begin{array}{r} 1\ 2 \\ \times\ \ 3 \\ \hline 6 \end{array}$	$\begin{array}{r} 1\ 2 \\ \times\ \ 3 \\ \hline 3\ 6 \end{array}$
	Multiply: $3 \times 2 = 6$ Place the produce in the *ones* column.	Multiply by 3 the 1 in the *tens* column: $3 \times 1 = 3$ Place your answer in the *tens* column.

Practice Exercise 43

Find each product. Work slowly and carefully, using the sample method just demonstrated.

1. $\begin{array}{r} 43 \\ \times\ 2 \\ \hline \end{array}$ 2. $\begin{array}{r} 12 \\ \times\ 4 \\ \hline \end{array}$ 3. $\begin{array}{r} 20 \\ \times\ 3 \\ \hline \end{array}$ 4. $\begin{array}{r} 31 \\ \times\ 3 \\ \hline \end{array}$ 5. $\begin{array}{r} 24 \\ \times\ 2 \\ \hline \end{array}$

6. $\begin{array}{r} 11 \\ \times\ 8 \\ \hline \end{array}$ 7. $\begin{array}{r} 33 \\ \times\ 2 \\ \hline \end{array}$ 8. $\begin{array}{r} 78 \\ \times\ 1 \\ \hline \end{array}$ 9. $\begin{array}{r} 10 \\ \times\ 7 \\ \hline \end{array}$ 10. $\begin{array}{r} 65 \\ \times\ 1 \\ \hline \end{array}$

Example

Mr. Vickerson earned $13 a day. After seven work days, how much had he earned?

Solution

The problem is to multiply 13×7.

Problem	Step 1	Step 2	Step 3
$\begin{array}{r} 13 \\ \times\ 7 \\ \hline \end{array}$	$\begin{array}{r} 1\ 3 \\ \times\ \ 7 \\ \hline 2\ 1 \end{array}$	$\begin{array}{r} 1\ 3 \\ \times\ \ 7 \\ \hline 2\ 1 \\ 7\ 0 \\ \hline \end{array}$	$\begin{array}{r} 1\ 3 \\ \times\ \ 7 \\ \hline 2\ 1 \\ +\ 7\ 0 \\ \hline 9\ 1 \end{array}$
	Multiply: $7 \times 3 = 21$ This is the first *partial* product.	Multiply: $7 \times 10 = 70$ This is the second *partial* product.	Add the partial products: $70 + 21 = 91$

Thus: $13 \times 7 = 91$.

Again we look for an easier method of multiplying. Recall that when we found a sum greater than 10 in any column we *carried* to the next column. This same method can be applied to multiplication also, as shown in the following example:

Problem	Step 1	Step 2
13 × 7	2 1 3 × 7 1	2 1 3 × 7 9 1
	Multiply the 7 times the ones number 3: 7 × 3 = 21 Write the 1 in the *ones* column and *carry* the 2 to the *tens* column.	Multiply the 7 times the tens number 1: 7 × 1 = 7 Add on the carried number 2; 7 + 2 = 9 Write the 9 in the *tens* column.

Thus: 13 × 7 = 91.

The same method can be used whenever you are multiplying by a single-digit number. It doesn't matter how large or small your multiplicand may be. Just remember that you multiply first and then add on, if necessary, the carried number.

Look over the next two examples before you attempt the practice exercise on page 109.

Problem	Step 1	Step 2	Step 3
125 × 6	3 1 2 5 × 6 0	1 3 1 2 5 × 6 5 0	1 3 1 2 5 × 6 7 5 0
	Multiply: 6 × 5 = 30 Place the 0 in the *ones* column and carry the 3 to the *tens* place.	Multiply: 6 × 2 = 12 Add the 3 you carried: 12 + 3 = 15 Write the 5 in the *tens* column and and carry the 1 to the *hundreds* column.	Multiply: 6 × 1 = 6 Add the 1 you carried: 6 + 1 = 7 Write the 7 in the *hundreds* column.

Thus: 125 × 6 = 750.

Problem	Step 1	Step 2	Step 3	Step 4
1,306 × 7	4 1306 × 7 2	4 1306 × 7 42	2 4 1306 × 7 142	2 4 1306 × 7 9142
	Multiply: 7 × 6 = 42 Write the 2 in the *ones* column and carry the 4.	Multiply: 7 × 0 = 0 Add the 4 you carried: 0 + 4 = 4 Write the 4 in the *tens* column.	Multiply: 7 × 3 = 21 Write the 1 in the *hundreds* column and carry the 2.	Multiply: 7 × 1 = 7 Add the 2 you carried: 7 + 2 = 9 Write the 9 in the *thousands* column.

Thus: 1,306 × 7 = 9,142.

Example

Let's return to the example in section 4.3. You are now able to help David figure his yearly earnings:

Solution

$$\begin{array}{r} 52 \\ \times\ \$4 \\ \hline \$208 \end{array} \qquad \begin{array}{r} 52 \\ \times\$1 \\ \hline \$52 \end{array} \qquad \begin{array}{r} 1 \\ 52 \\ \times\ \$5 \\ \hline \$260 \end{array}$$

$$\begin{array}{r} \$208 \\ 52 \\ 260 \\ \hline \$520 \end{array}$$ his total earnings for the year

Practice Exercise 44

Find the product for each of the following examples. The first five examples have been partially done for you.

1. $\begin{array}{r} 4 \\ 99 \\ \times\ 5 \\ \hline 5 \end{array}$ 2. $\begin{array}{r} 2 \\ 77 \\ \times\ 4 \\ \hline 8 \end{array}$ 3. $\begin{array}{r} 4 \\ 38 \\ \times\ 6 \\ \hline 8 \end{array}$ 4. $\begin{array}{r} 2 \\ 58 \\ \times\ 3 \\ \hline 4 \end{array}$ 5. $\begin{array}{r} 1 \\ 42 \\ \times\ 9 \\ \hline 8 \end{array}$

6. $\begin{array}{r} 282 \\ \times\ 3 \\ \hline \end{array}$ 7. $\begin{array}{r} 129 \\ \times\ 3 \\ \hline \end{array}$ 8. $\begin{array}{r} 245 \\ \times\ 7 \\ \hline \end{array}$ 9. $\begin{array}{r} 307 \\ \times\ 8 \\ \hline \end{array}$ 10. $\begin{array}{r} 773 \\ \times\ 7 \\ \hline \end{array}$

11. $\begin{array}{r} 1,038 \\ \times\ 5 \\ \hline \end{array}$ 12. $\begin{array}{r} 976 \\ \times\ 6 \\ \hline \end{array}$ 13. $\begin{array}{r} 540 \\ \times\ 7 \\ \hline \end{array}$ 14. $\begin{array}{r} 165 \\ \times\ 5 \\ \hline \end{array}$ 15. $\begin{array}{r} 2,389 \\ \times\ 8 \\ \hline \end{array}$

Review Of Important Ideas

Some important ideas covered in Chapter 4 so far are:

Multiplication is "repeated addition."

The symbol × means to multiply.

Numbers are multiplied by pairs only.

The order of the multiplicand and multiplier doesn't affect the product.

For a product greater than 10, carry the tens number to the next column and add it to the next product.

4.6 Word Problems

Review section 2.7 in Chapter 2 for general instructions needed to solve any word problem. The following are specific instructions for solving a multiplication example.

Key Words Or Phrases For Problems Solved By Multiplying

There are certain key words or phrases that indicate the problem will be solved by multiplication. Recall that multiplication means repeated *addition,* so all those key words can be used again. You must recognize that the problem contains *repeated* additions and that you will obtain a faster result by multiplication.

1. *Product* — "Find the *product* of 46 times 9."
2. *Times* — "Three *times* as many adults attended as children."
3. *How many* and *how much* — "*How many* apartments are there in the building?" or "*How much* did it cost?"

Practice Exercise 45

Search out the key words or phrases and then do each problem.

1. For the past 6 days in a row, I played tennis. If the cost is $4 a day, how much did it cost?

2. There are 6 floors in my apartment building and 8 apartments on each floor. How many apartments are there in my building?

3. A family of 4 went to a movie. What was the cost of admission if it was $2 for each person?

4. A car travels 18 mi. on a gallon of gasoline. How many miles can it travel on 8 gal. of gasoline?

5. Find the product of 46 times 9.

6. If the charge of drilling a well is $4 a foot, how much would it cost to drill a well 35 ft. deep?

7. If 1,063 people visited the shrine each day, how many people visited the shrine in 7 days?

8. A restaurant has 246 tables. If each table seats 4 people and the restaurant is filled at 7:00 P.M., how many people are seated at the tables?

9. Last week 163 children went to the circus on Thursday night. Three times as many children went on Saturday afternoon. How many children were there on Saturday afternoon?

10. If 4,873 people crossed the border from Canada to the United States each day, how many people would have crossed from Canada to the United States in 4 days?

4.7 Multiplying By A Number Having Two Or More Digits

Multiplication of this type is more involved but not more difficult. You will continue using all the laws you have learned concerning the multiplication of whole numbers.

For example, to find the product of 31 times 18, you could add 18 thirty-one times or add 31 eighteen times. You know that multiplication is an easier way of doing the example, so let's multiply 31 × 18.

Problem	Step 1	Step 2	Step 3
31 ×18	3 1 × 1 8 2 4 8	3 1 × 1 8 2 4 8 3 1 0	3 1 × 1 8 2 4 8 + 3 1 0 5 5 8
	Multiply: 8 × 31 = 248	Multiply: 10 × 31 = 310 (Remember the 1 in the *tens* place is 10.)	Add the partial products: 248 + 310 = 558

Thus: 31 × 18 = 558.

Notice that 31 × 10 = 310. An earlier example on page 107 showed that 7 × 10 = 70. Do you know the product of 15 × 10? It is 150. Multiplying by 10 can be accomplished by simply adding a *zero* to the right of the multiplicand. Look at these examples:

$$\begin{array}{r} 36 \\ \times 10 \\ \hline 36\underline{0} \end{array} \qquad \begin{array}{r} 493 \\ \times\ 10 \\ \hline 4{,}93\underline{0} \end{array} \qquad \begin{array}{r} 1{,}063 \\ \times\quad 10 \\ \hline 10{,}63\underline{0} \end{array}$$

To multiply by 100, multiply the same way as you did by 10, except you must add *two zeros* to the right of the multiplicand.

$$\begin{array}{r} 36 \\ \times 100 \\ \hline 36\underline{00} \end{array} \qquad \begin{array}{r} 493 \\ \times 100 \\ \hline 49{,}3\underline{00} \end{array} \qquad \begin{array}{r} 1{,}063 \\ \times\quad 100 \\ \hline 106{,}3\underline{00} \end{array}$$

The same method is used for any multiplication of 10; 100; 1,000; 10,000; etc. For a multiplier of 20 you proceed as if you multiplied by a 2 and then a 10. For example:

Problem	Step 1	Step 2
36 ×20	36 ×20 72	36 ×20 720
	Multiply: 36 × 2 = 72	Multiplying by 10 results in adding a *zero* to the 72.

Thus: 36 × 20 = 720.

Let's try this short practice exercise first before continuing. Do it slowly and carefully.

Practice Exercise 46

Multiply. The first three examples have been partially completed.

1. 14 × 10 = 14______

2. 56 × 100 = 56______

3. 23 × 20 = 46______

4. 124 × 100 =

5. 36 × 30 =

6. 51 × 70 =

7. 306 × 90 =

8. 300 × 20 =

9. 123 × 10 =

10. 500 × 200 =

Let's turn to another example.

Problem	Step 1	Step 2	Step 3
47 ×39	4 7 × 3 9 4 2 3	4 7 × 3 9 4 2 3 1 4 1 0	4 7 × 3 9 4 2 3 + 1 4 1 0 1,8 3 3
	Multiply: 9 × 47 = 423	Multiply: 30 × 47 = 1,410	Add the partial products: 423 + 1,410 = 1,833

Problem	Step 1	Step 2	Step 3
146 × 73	1 4 6 × 7 3 4 3 8	1 4 6 × 7 3 4 3 8 1 0 2 2 0	1 4 6 × 7 3 4 3 8 + 1 0 2 2 0 1 0 6 5 8
	Multiply: 3 × 146 = 438	Multiply: 70 × 146 = 10,220	Add the partial products: 438 + 10,220 = 10,658

Our goal is to do a multiplication example as simply and as speedily as possible. Look at the same problem again, but this time let's do it in our heads.

It would look like this	*It would sound like this*	
146	*Step 1:*	3 × 6 = 18. Write the 8 and carry the 1.
× 73	*Step 2:*	3 × 4 = 12. Add the 1 you carried: 12 + 1 = 13. Write the 3 and carry the 1.
438	*Step 3:*	3 × 1 = 3. Add the 1 you carried: 3 + 1 = 4. Write the 4.
1,022	*Step 4:*	7 × 6 = 42. Write the 2 directly under the number you multiplied by (7) and carry the 4.
10,658	*Step 5:*	7 × 4 = 28. Add the 4 you carried: 28 + 4 = 32. Write the 2 and carry the 3.
	Step 6:	7 × 1 = 7. Add the 3 you carried: 7 + 3 = 10. Write the 10.
	Step 7:	Add to find the final answer.

Practice will help you achieve better results. All of the above will help you to understand the process involved in the multiplication of large numbers. If you find you are running into difficulties doing the next exercise, go back to the illustrated examples in this sections, reread them, and then try the practice exercise again.

Practice Exercise 47

Multiply.

1. 58 ×43	2. 39 ×28	3. 56 ×35	4. 92 ×87	5. 61 ×47
6. 830 × 34	7. 605 × 12	8. 391 ×109	9. 172 × 65	10. 989 × 81
11. 1,006 × 39	12. 650 × 45	13. 348 ×290	14. 6,204 × 109	15. 9,992 × 88

4.8 Multiplication Applied To Measuring

Area

Example

The Reed family had an L-shaped living room with dimensions as illustrated in the diagram below. They decided to cover the floor with wall-to-wall carpeting. They found that carpeting cost $4 a square yard in a special half-price sale. How much did it cost to have the room carpeted?

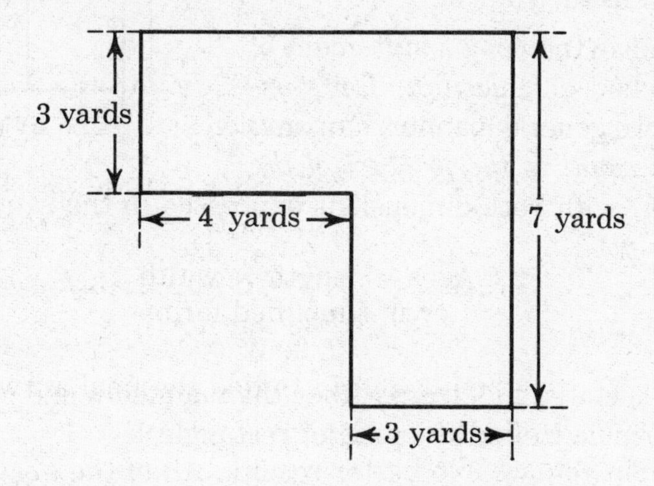

Solution

If you wish to find the measurement of the interior of a figure, then you must find its *area*. To find the area, you must know the number of square units required to cover it or the number of square units the region contains. In our problem, you would use a square yard like this to find the area.

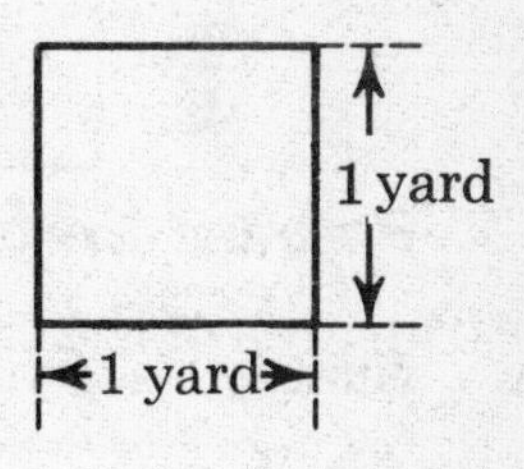

The L-shaped living room can be subdivided into two rectangles, as in the following diagram:

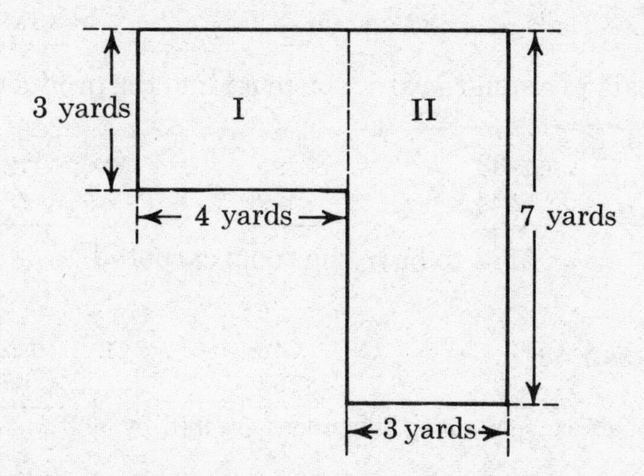

The problem is then simplified to finding the area of the two rectangles marked I and II above.

Thus, the area is 12 sq. yd., since this is what was needed to cover the interior of the rectangle. This could also be done by finding the product of the length and the width of the rectangle.

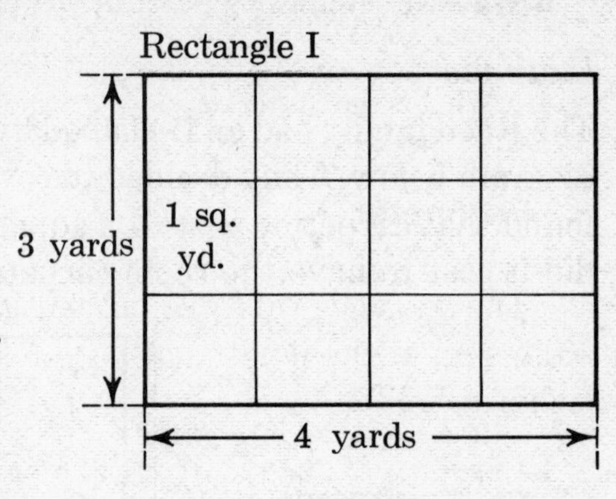

$$\text{Area} = 3 \text{ yd.} \times 4 \text{ yd.} \quad \text{or}$$
$$A = 12 \text{ sq. yd.}$$

Recall that earlier in the book a statement of this type was called an *equation.* Let's express this in a more general manner: *The area of a rectangle is equal to the product of the length and width,* with both dimensions expressed in the same unit of measurement, or

$$\text{Area} = \text{length} \times \text{width}$$
or in simplified form
$$A = L \times W$$

This equation is stated in letters rather than numbers and is called a *formula.* Let's use this formula to find the area of rectangle II.

Rectangle II

7 yards

3 yards

$$\text{Area} = \text{length} \times \text{width}$$
$$A = L \times W$$
$$A = 7 \text{ yd.} \times 3 \text{ yd.}$$
$$A = 21 \text{ sq. yd.}$$

The area of the L-shaped living room is the sum of the areas of the two rectangles.

$$\text{Area} = \text{rectangle I} + \text{rectangle II}$$
$$A = 12 \text{ sq. yd.} + 21 \text{ sq. yd.}$$
$$A = 33 \text{ sq. yd.}$$

Since carpeting costs \$4 a square yard, you must find the product of \$4 × 33 to know the total cost.

$$\begin{array}{r} 33 \\ \times \$4 \\ \hline \$132 \end{array}$$ to have the room carpeted

Practice Exercise 48

1. Find the area of a rectangle whose dimensions are 9 ft. by 36 ft.

2. What is the area of a room whose dimensions are 5 yd. by 6 yd.?

3. A sheet of drawing paper measures 14 in. by 18 in. What is its area?

4. What is the cost of scraping and shellacking an 8 ft. × 15 ft. floor if it costs $1 a square foot?

Perimeter

The *perimeter* of a geometric figure, as mentioned in section 2.14 of chapter 2, is the sum of the lengths of its sides. The perimeter of a *rectangle* 8 ft. by 15 ft. is shown below.

15′

8′ 8′

15′

Perimeter = 8 ft. + 8 ft. + 15 ft. + 15 ft.
P = 16 ft. + 30 ft.
P = 46 ft.

This statement is another example of an *equation.* Suppose you had a rectangle which did not have any particular dimensions. How can you express its perimeter?

L

W W

L

Perimeter = $L + L + W + W$ where
L = length and W = width.

This is a formula for finding the perimeter of the rectangle since it is a rule expressed in terms of letters rather than numbers. The repeated addition of $L + L$ can be written as $2 \times L$ and $W + W$ as $2 \times W$. In mathematics you can drop the *times* sign between the letter and number, and write $2L$ and $2W$. Even though the multiplication sign is not written it is understood to be there.

Thus, $P = L + L + W + W$ becomes $P = 2L + 2W$. This is the formula we can use to find the perimeter of any rectangle.

Example

Find the perimeter of a rectangle whose dimensions are 8 ft. by 15 ft.

Solution

8′

15′

$P = 2L + 2W$
$P = 2 \times 15$ ft. + 2×8 ft.
$P = 30$ ft. + 16 ft.
$P = 46$ ft.

How can we develop a formula for the perimeter of an equilateral triangle? An equilateral triangle is a triangle which has three equal sides. Let's represent the length of each side by the letter s.

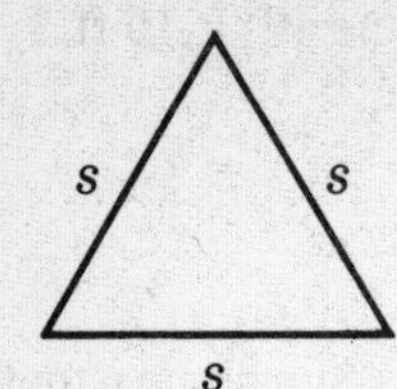

Then, $P = s + s + s$

$P = 3 \times s$ (repeated addition)

$P = 3s$ (dropping the *times* sign)

Example

Find the perimeter of a square whose side measures 6 in.

Solution

The formula for the perimeter of a square is

$$P = 4s$$
$$P = 4 \times 6 \text{ in.}$$
$$P = 24 \text{ in.}$$

This step is called *substitution.* It means we replaced the letter s by the numerical value 6 in.

Example

Each side of a pentagon is 1,500 ft. What is its perimeter?

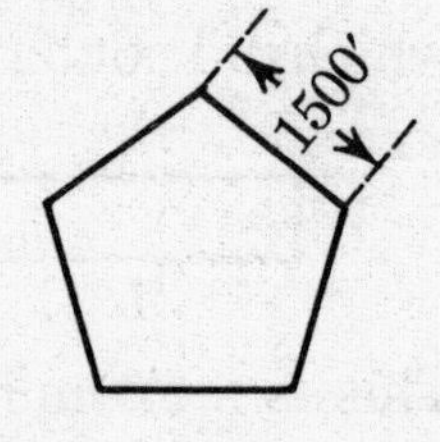

Solution

A pentagon is a five-sided closed plane geometric figure as shown above. Each side measures 1,500 ft. Thus, the formula for the perimeter is

$$P = s + s + s + s + s$$
$$P = 5s$$
$$P = 5 \times s$$
$$P = 5 \times 1500 \text{ ft.}$$
$$P = 7500 \text{ ft.}$$

Practice Exercise 49

Using formulas, solve the following problems:

1. Using the formula $P = 3s$, find the perimeter if the value of s is 34 ft.

2. How many feet of split fencing are needed to enclose a rectangular-shaped garden with dimensions of 15 ft. by 12 ft.?

3. Find the perimeter of a square if each side measures 16 ft.

4. Using the formula $P = 2L + 2W$, what is the value of P if $L = 32$ ft. and $W = 10$ ft.?

5. Using the formula $P = 12x$, what is the value of P if $x = 17$ in.?

6. Write a formula for the perimeter of this triangle. (Remember: The perimeter is the sum of all the sides of the figure, so $P = a + a + b$.) Complete the formula, writing it in its simplest form as was shown earlier.

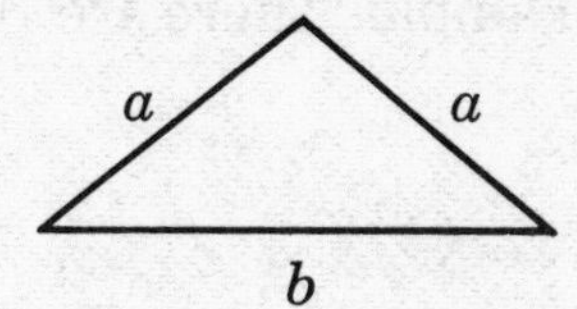

7. Farmer Appleby owns 62 ft. of wire and wants to string it in the shape of a triangle. Two of the sides of the triangle must measure 23 ft. each. What should be the length of the third side?

Terms You Should Remember

Area The measure of space within an enclosed region.

Perimeter The sum of the lengths of the sides of a closed geometric figure.

Square inch A square whose sides are 1 in.

Square foot A square whose sides are 1 ft.

Square yard A square whose sides are 1 yd.

4.9 More Difficult Word Problems Requiring Multiplication

Example

A shopkeeper bought 76 dolls. Each doll cost \$13. How much did he pay for all of them?

Solution

This is a problem in repeated addition or multiplication. You want to add \$13 a total of 76 times or 76 × \$13. Since 76 × \$13 is the same as \$13 × 76 choose the number you want as your multiplier. Choosing 13 as the multiplier makes the example look like this:

76	*Step 1:*	$3 \times 6 = 18$. Write the 8 and carry the 1.
$\times$\$13	*Step 2:*	$3 \times 7 = 21$. Add the 1 you carried; $21 + 1 = 22$.
228	*Step 3:*	$1 \times 6 = 6$. Write the 6 in the column exactly under the 1.
76		
\$988	*Step 4:*	$1 \times 7 = 7$. Write the 7 to the left of the 6. Then add to find the final answer.

Before beginning this practice exercise, reread Section 4.6 to find the key words that indicate a multiplication example. There are 10 problems in this exercise. Try for a perfect score.

Practice Exercise 50

1. A sewing machine operator makes 125 articles daily. At the end of a five-day work week, how many articles has she completed?

2. A candy store owner sold 652 pieces of candy. Each piece of candy costs 15¢. How many cents does he collect if he sells all his candy?

3. A shipping clerk mails 15 cartons to each of 740 customers. What is the total number of cartons he mails?

4. How much does a man earn in 14 weeks if his weekly earnings are \$135?

5. A car-leasing company rents 1,435 cars monthly. How many cars does it rent by the end of one year? (Remember, 12 months = 1 year.)

6. A union local has 2,432 members. If each member pays \$52 dues a year, how much money does the local collect?

7. At Angie's Pizza Parlor, each slice of pizza costs 45¢. If each pie is cut into 8 slices, how much money (in cents) does the restaurant get for two pies?

8. On each boat trip, a "dragger" returns with 3,800 lb. of fish. How many pounds of fish will he bring in after 61 trips?

9. Sam purchases 200 leather wallets at $11 each. If he sells all of them, collecting $3,200, how much profit does he make?

10. The rent for a 3-room apartment is $225 a month. How many dollars do you pay for rent in 3 years?

Review Of Important Ideas

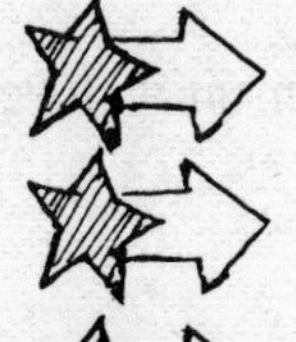

An equation is a sentence of equality.

A formula is a rule stated in letters rather than numbers.

The area of a rectangle = length × width.

The perimeter of a geometric figure is equal to the sum of its sides. All dimensions must be in the same unit of measurement.

Replacing a letter with a number in a formula is called *substitution.*

Let's See How Your Skill In Multiplication Has Improved

This chapter covered multiplication of whole numbers. Many different types of problems were included and your skill in computation has probably improved since the pretest. Here is a way to find out, since the posttest will let you check on your own understanding.

Posttest 4

Write your answers in the spaces provided.

1.	2.	3.	4.	5.
8 ×6	43 × 2	313 × 3	64 × 2	304 × 5
6.	**7.**	**8.**	**9.**	**10.**
440 × 2	24 ×13	63 ×26	87 ×56	402 × 86

11. $\begin{array}{r} 610 \\ \times\ 80 \\ \hline \end{array}$

12. $\begin{array}{r} 600 \\ \times 300 \\ \hline \end{array}$

13. $\begin{array}{r} 487 \\ \times\ 89 \\ \hline \end{array}$

14. Find the product of 4 × 2 on the number line.

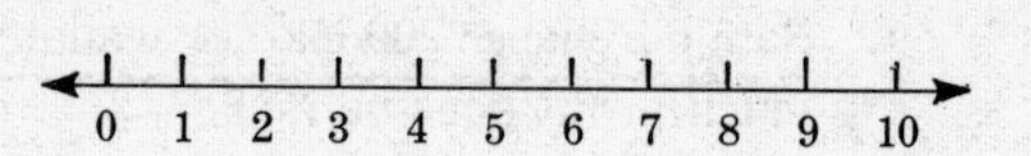

15. How many cents do you save in a Christmas Club savings account if you save 25¢ a week for 50 weeks?

16. Two cans of beer cost 43¢. How many cents do 6 cans of beer cost?

17. What is the total cost of 4 lb. of lamb stew at 89¢ a pound and 2 lb. of chopped meat at 109¢ a pound?

18. Your car travels 12 mi. for each gallon of gasoline. How many miles can you travel on 12 gal.?

19. If a stadium seats 54,260 persons, how many people would have attended 23 events if it was filled each time?

20. A rectangular-shaped garden has dimensions of 26 ft. by 45 ft. What is the area of the garden?

21. Find the product of 1,723 and 391.

22. Find the value of P in the formula $P = 6s$, if $s = 39$ ft.

ANSWERS AND EXPLANATIONS TO POSTTEST 4

1. 48	2. 86	3. 939	4. 128	5. 1,520
6. 880	7. 312	8. 1,638	9. 4,872	10. 34,572
11. 48,800	12. 180,000	13. 43,343		

14. 

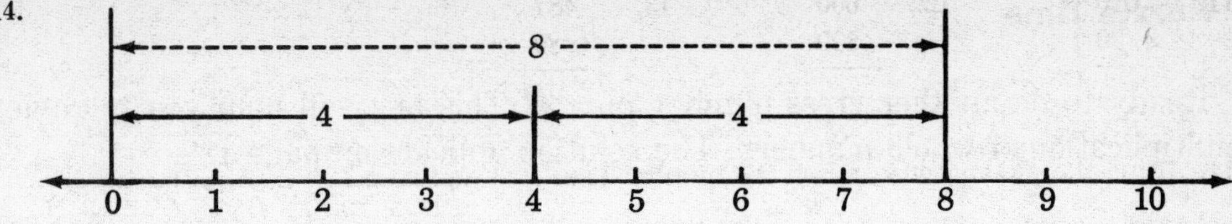

15.
25¢
× 50
1250¢

16.
43¢
× 3
129¢

17.

89¢	109¢	356¢
× 4	2	+218¢
356¢	218¢	574¢

18.
12
×12
24
12
144 mi.

19.
54,260
×23
162 780
1 085 20
1,247,980

20.
26′
45′

$A = L \times W$
$A = 45$ ft. × 26 ft.
$A = 1{,}170$ sq. ft.

21.
1,723
× 391
1 723
155 07
516 9
673,693

22.
$P = 6s$
$P = 6 \times 39$ ft.
$P = 234$ ft.

A Score of	*Means That You*
19–22	Did very well. You can move to Chapter 5.
17–18	Know this material except for a few points. Reread the sections about the ones you missed.
14–16	Need to check carefully on the sections you missed.
0–13	Need to review the chapter again to refresh your memory and improve your skills.

Questions	*Are Found in Section*
14	4.2
1	4.4
2–6	4.5
16, 17	4.6
7–13	4.7
20, 22	4.8
15, 18, 19, 21	4.9

Puzzle Time

Ready to do another cross-number puzzle? This one will help you to review multiplication of whole numbers. The solution appears on page 128.

ACROSS

2. $67 \times 9 =$
3. $38 \times 9 =$
6. $342 \times 5 =$
9. $479 \times 6 =$
11. $4 \times 7 =$
12. $6 \times 9 =$
14. $195 \times 5 =$
17. $28 \times 3 =$
19. $9 \times 7 =$
20. $163 \times 5 =$
22. $5 \times 5 =$
24. $4 \times 3 =$
26. $78 \times 54 =$
30. $147 \times 15 =$
31. $48 \times 18 =$
32. $14 \times 27 =$

DOWN

1. $12 \times 11 =$
2. $4 \times 15 =$
4. $8 \times 6 =$
5. $9 \times 3 =$
7. $24 \times 3 =$
8. $7 \times 27 =$
10. $9 \times 5 =$
13. $27 \times 18 =$
15. $54 \times 14 =$
16. $29 \times 19 =$
18. $24 \times 18 =$
21. $32 \times 16 =$
23. $6 \times 9 =$
25. $4 \times 5 =$
27. $7 \times 4 =$
28. $8 \times 2 =$
29. $12 \times 20 =$
30. $14 \times 2 =$

ANSWERS FOR CHAPTER 4

PRETEST 6

1. 35
2. 68
3. 848
4. 219
5. 2,412
6. 660
7. 420
8. 1,462
9. 2,632
10. 28,458
11. 35,700
12. 21,090
13. 100,000
14. 37,044
15. 258¢

16.

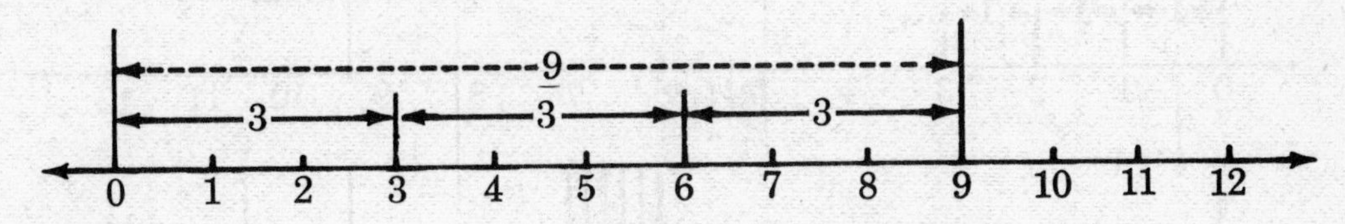

17. 1,250¢
18. 340¢ + 130¢ = 470¢
19. 144 mi.
20. $33,024
21. 78 sq. ft.
22. P = 44 in.

PRACTICE EXERCISE 39

1.

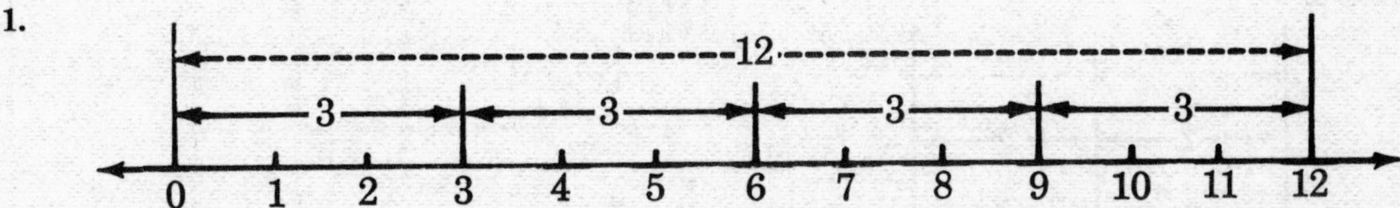

2.

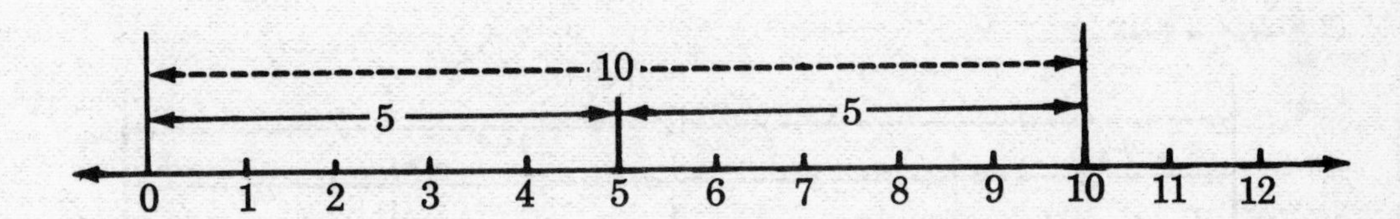

3.

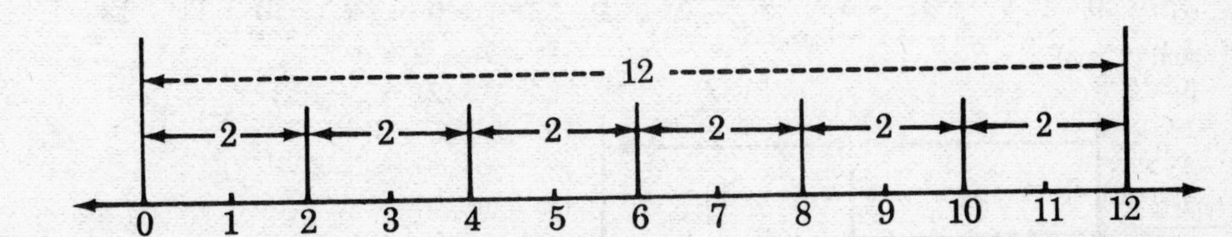

PRACTICE EXERCISE 40

1. (1 × 4)

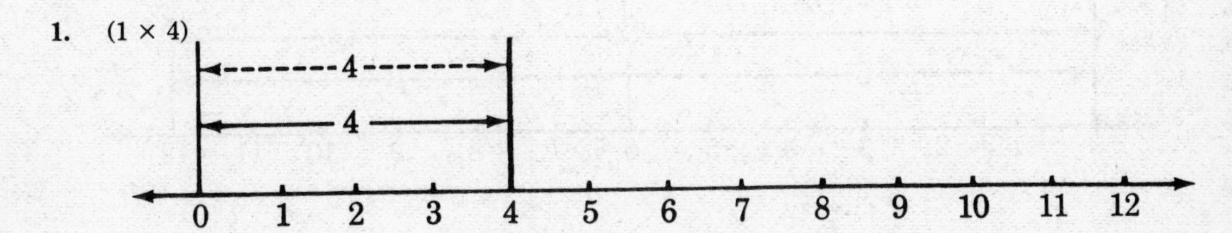

$3 \times (1 \times 4) = 12$

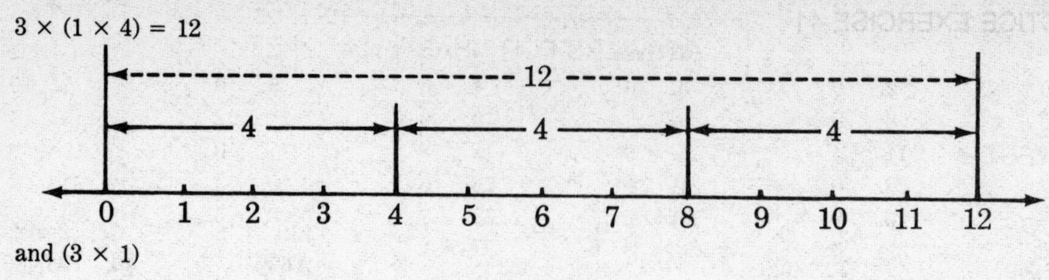

and (3×1)

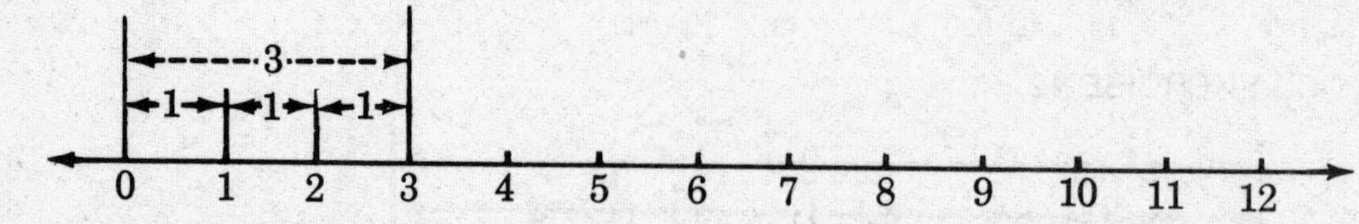

$(3 \times 1) \times 4 = 12$

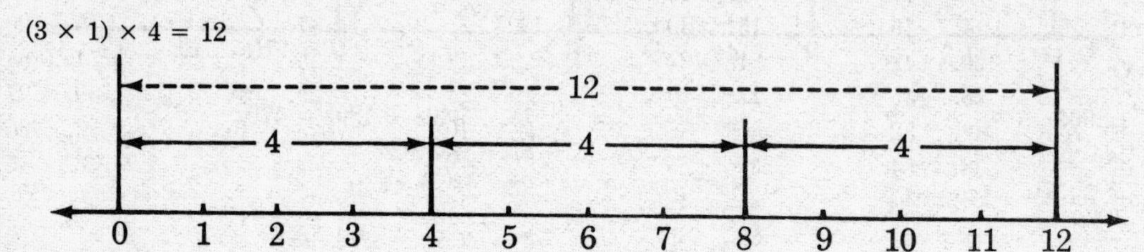

2. (2×2)

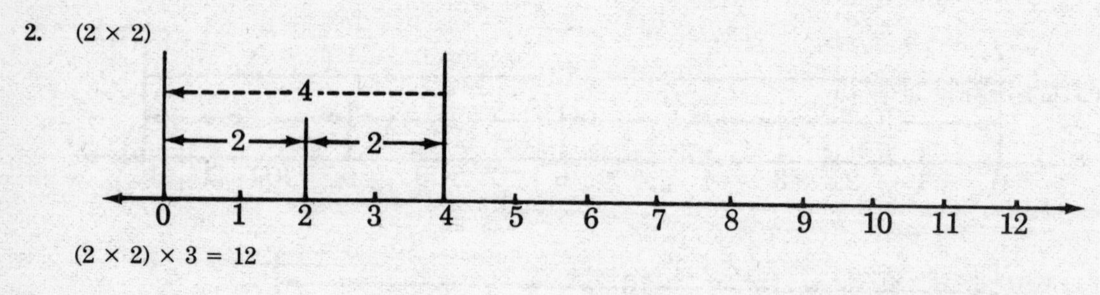

$(2 \times 2) \times 3 = 12$

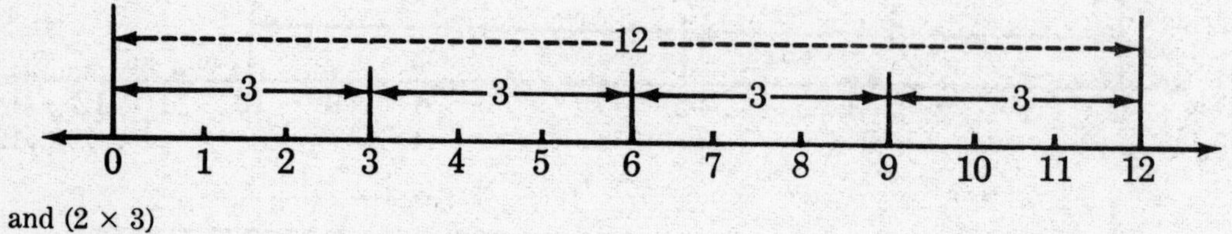

and (2×3)

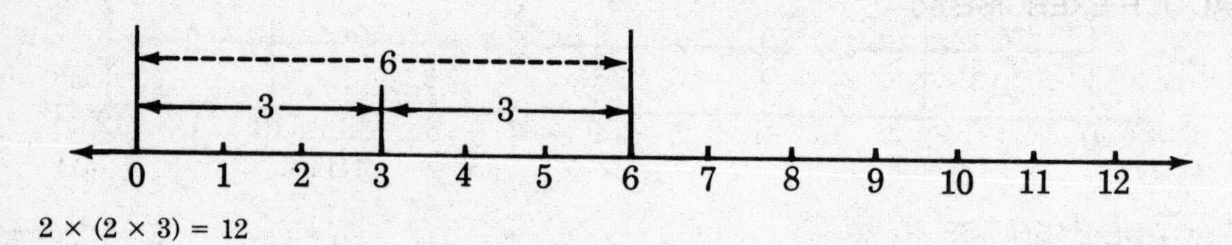

$2 \times (2 \times 3) = 12$

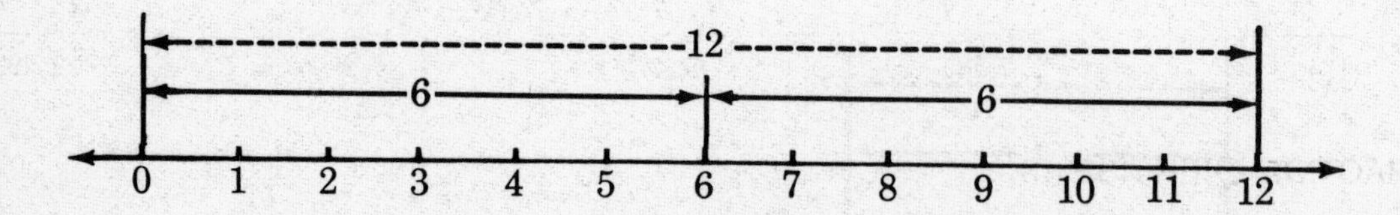

PRACTICE EXERCISE 41

1. 0 **2.** 0 **3.** 0 **4.** 0 **5.** 0 **6.** 0
7. 0 **8.** 0 **9.** 0 **10.** 0 **11.** 2 **12.** 1
13. 0 **14.** 4 **15.** 9 **16.** 5 **17.** 0 **18.** 0
19. 0 **20.** 0 **21.** 0 **22.** 0 **23.** 3 **24.** 8
25. 2 **26.** 9 **27.** 0 **28.** 3 **29.** 5 **30.** 0
31. 6 **32.** 4 **33.** 8 **34.** 6 **35.** 7 **36.** 7

PRACTICE EXERCISE 42

1. 30 **2.** 18 **3.** 14 **4.** 45 **5.** 56 **6.** 24
7. 42 **8.** 54 **9.** 48 **10.** 49 **11.** 63 **12.** 63
13. 40 **14.** 16 **15.** 30 **16.** 21 **17.** 36 **18.** 20
19. 24 **20.** 42 **21.** 21 **22.** 28 **23.** 81 **24.** 64
25. 27 **26.** 35 **27.** 40 **28.** 24 **29.** 36 **30.** 27
31. 16 **32.** 32 **33.** 36 **34.** 10 **35.** 15 **36.** 56
37. 28 **38.** 54 **39.** 25 **40.** 18 **41.** 72 **42.** 15
43. 72 **44.** 14 **45.** 48 **46.** 4 **47.** 32 **48.** 6
49. 6 **50.** 45 **51.** 8 **52.** 9 **53.** 20 **54.** 10
55. 18 **56.** 24 **57.** 16 **58.** 35 **59.** 12 **60.** 12
61. 12 **62.** 12 **63.** 18 **64.** 8

PRACTICE EXERCISE 43

1. 86 **2.** 48 **3.** 60 **4.** 93 **5.** 48
6. 88 **7.** 66 **8.** 78 **9.** 70 **10.** 65

PRACTICE EXERCISE 44

1. 495 **2.** 308 **3.** 228 **4.** 174 **5.** 378
6. 846 **7.** 387 **8.** 1,715 **9.** 2,456 **10.** 5,411
11. 5,190 **12.** 5,856 **13.** 3,780 **14.** 825 **15.** 19,112

PRACTICE EXERCISE 45

1. $4 × 6 = $24
2. 8 × 6 = 48
3. $2 × 4 = $8
4. 18 × 8 = 144 mi.
5. 46 × 9 = 414
6. 35 × $4 = $140
7. 1,063 × 7 = 7,441
8. 246 × 4 = 984
9. 163 × 3 = 489
10. 4,873 × 4 = 19,492

PRACTICE EXERCISE 46

1. 140 **2.** 5,600 **3.** 460 **4.** 12,400 **5.** 1,080
6. 3,570 **7.** 27,540 **8.** 6,000 **9.** 1,230 **10.** 100,000

PRACTICE EXERCISE 47

1.	2,494	**2.**	1,092	**3.**	1,960	**4.**	8,004	**5.**	2,867
6.	28,220	**7.**	7,260	**8.**	42,619	**9.**	11,180	**10.**	80,109
11.	39,234	**12.**	29,250	**13.**	100,920	**14.**	676,236	**15.**	879,296

PRACTICE EXERCISE 48

1. 324 sq. ft. **2.** 30 sq. yd. **3.** 252 sq. in. **4.** $120

PRACTICE EXERCISE 49

1. $P = 3s$
$P = 3 \times 34$ ft.
$P = 102$ ft.

2. $P = 2L + 2W$
$P = 2 \times 15$ ft. $+ 2 \times 12$ ft.
$P = 30$ ft. $+ 24$ ft.
$P = 54$ ft.

3. $P = 4s$
$P = 4 \times s$
$P = 4 \times 16$ ft.
$P = 64$ ft.

4. $P = 2L + 2W$
$P = 2 \times 32$ ft. $+ 2 \times 10$ ft.
$P = 64$ ft. $+ 20$ ft.
$P = 84$ ft.

5. $P = 12x$
$P = 12 \times 17$ in.
$P = 204$ in.

6. $P = a + a + b$
$P = 2 \times a + b$
$P = 2a + b$

7. $P = 2a + b$
62 ft. $= 2 \times 23$ ft. $+ b$
62 ft. $= 46$ ft. $+ b$
16 ft. $= b$

PRACTICE EXERCISE 50

1.	625	**2.**	9,780¢	**3.**	11,100	**4.**	$1,890	**5.**	17,220
6.	$126,464	**7.**	720¢	**8.**	231,800 lb.	**9.**	$1,000	**10.**	$8,100

CROSS-NUMBER PUZZLE SOLUTION

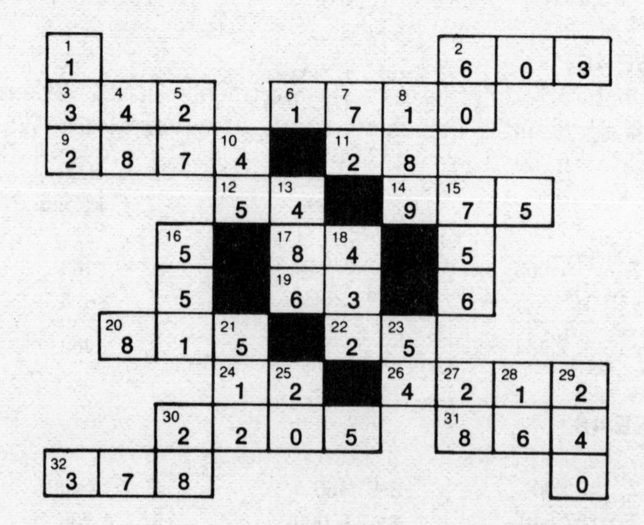

You have done well. With the knowledge you have gathered in these chapters, you will find that working with numbers helps you in everyday situations. Chapter 5 covers division and will complete the four basic operations with numbers.

Practice your skills as often as you can. They are only tools that develop further learning.

Before you start Chapter 5, think about this:

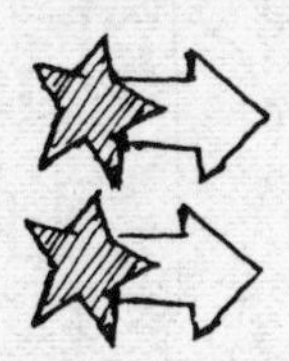

Reading in mathematics improves as you know more words or terms.

Reading of mathematics requires close attention to relationships.

1. Which ideas are connected or related?
2. How does one fact lead to or cause another?
3. Which idea follows next in order of sequence?

Hold It!

This is a general review of some of the skills you have learned up to this point. You are making good progress if you get all the answers to these problems correct.

1. $\begin{array}{r} 68 \\ 137 \\ +\ 97 \\ \hline \end{array}$ **2.** $\begin{array}{r} 4{,}352 \\ -1{,}527 \\ \hline \end{array}$ **3.** $\begin{array}{r} 56 \\ \times\ 7 \\ \hline \end{array}$ **4.** $\begin{array}{r} 932 \\ 473 \\ +948 \\ \hline \end{array}$ **5.** $\begin{array}{r} 5{,}000 \\ -2{,}733 \\ \hline \end{array}$

6. $\begin{array}{r} 402 \\ \times\ 37 \\ \hline \end{array}$ **7.** $\begin{array}{r} 7 \\ 6 \\ 5 \\ 8 \\ +9 \\ \hline \end{array}$ **8.** $\begin{array}{r} 973 \\ -691 \\ \hline \end{array}$ **9.** $\begin{array}{r} 38 \\ \times 19 \\ \hline \end{array}$

10. At a major automobile factory there are 26 departments with 24 workers each, 18 departments with 28 workers each, and 25 departments with 31 workers each. What is the total working force?

ANSWERS TO "HOLD IT!"

1. 302 **2.** 2,825 **3.** 392 **4.** 2,353 **5.** 2,267

6. 14,874 **7.** 35 **8.** 282 **9.** 722

10. $\begin{array}{r} 26 \\ \times 24 \\ \hline 624 \end{array}$ $\begin{array}{r} 28 \\ \times 18 \\ \hline 504 \end{array}$ $\begin{array}{r} 31 \\ \times 25 \\ \hline 775 \end{array}$ $\begin{array}{r} 624 \\ 504 \\ +775 \\ \hline 1{,}903 \end{array}$

If six cans of X-Cola cost $1.80, what is the cost of one can? If a 12-foot board of lumber costs $1.92, what is the cost of one foot? These are two of many instances where *division* is needed to find the answer. *Division* is our fourth and final operation associated with whole numbers.

How well you can do on this topic can be checked with the pretest which follows.

See What You Know And Remember — Pretest 5

Work these exercises carefully, doing as many problems as you can. Write each answer in the space provided.

1. $6\overline{)42}$

2. $69 \div 3 =$

3. $2\overline{)426}$

4. $5\overline{)505}$

5. $4\overline{)208}$

6. $8\overline{)880}$

7. $\frac{65}{9} =$

8. $513 \div 3 =$

9. $4\overline{)824}$

10. $7\overline{)896}$

11. What would be the share for each person if we divided \$63 among 9 people?

12. Illustrate $6 \div 2$ on the number line below.

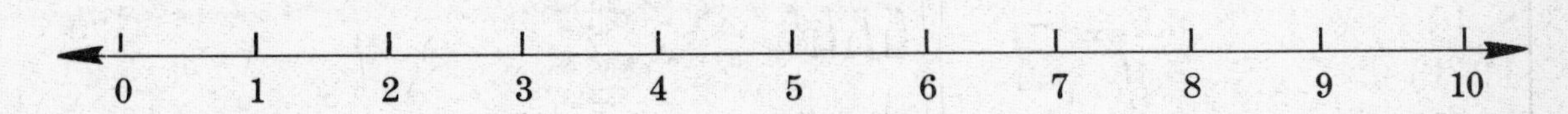

13. Seven people buy an equal share in the purchase of a lottery ticket. If they win \$56,049, how much does each one receive?

14. $85 \div 17 =$ 15. $24\overline{)504}$ 16. $32\overline{)1{,}664}$ 17. $42\overline{)6{,}066}$

18. Find the average height (using the mean, median, and mode methods) of five women whose individual heights are: 62 in., 65 in., 58 in., 62 in., and 68 in.

19. The area of a rectangle contains 132 sq. ft. If the length is 22 ft., find the width.

?′ Area = 132 sq. ft.

22′

20. Mr. Smith showed a yearly income of \$8,575 on his federal income tax return last year. This represented his earnings for 35 weeks during the year. How much did he earn each week?

Now turn to the end of the chapter to check your answers. Add up all that you had correct. Count by the number of separate answers, not by the numbers of questions. In the pretest, there were 20 questions, but 22 separate answers.

A Score of	*Means That You*
20–22	Did very well. You can move to Chapter 6 in Volume 2.
17–19	Know this material except for a few points. Read the sections about the ones you missed.
14–16	Need to check carefully on the sections you missed.
0–13	Need to work with the chapter to refresh your memory and improve your skills.

Questions	*Are Covered in Section*
1	5.3
12	5.4
2–4	5.5
5	5.6
6, 9	5.7
7	5.8
8, 10	5.9
11, 13	5.10
14–17	5.11
20	5.12
18	5.13
19	5.14

5.1 What Is Division?

If you recall, it was stated that subtraction is the inverse of addition. *Division* is the inverse of multiplication. Consider the following problem:

$$3 \overline{)\,12}$$

The problem asks you to divide 12 into 3 equal parts or $3 \times ? = 12$. Since $3 \times 4 = 12$, then

$$\begin{array}{r} 4 \\ 3 \overline{)\,12} \end{array}$$

We picture $3 \overline{)\,12}$ as 12 items divided into 3 groups.

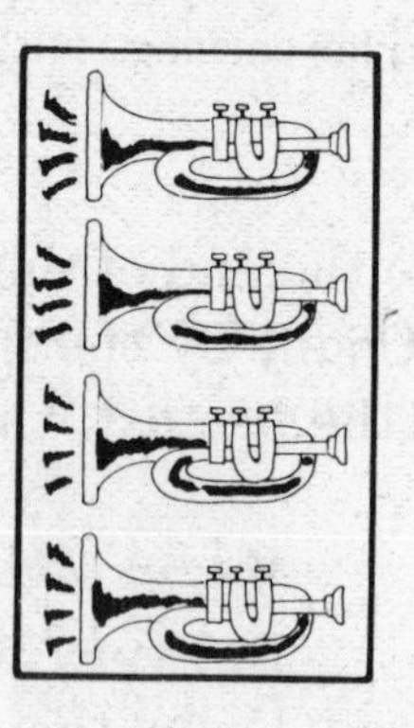

There are 4 items in each group. Thus:

$$\begin{array}{r} 4 \\ 3 \overline{)\,12} \end{array}$$

Multiplication was an operation of repeated additions. Division is an operation of repeated subtractions. Using the same example, $3\overline{)12}$, you subtract 3 from 12 continually until you can't subtract any more. The number of times you were able to subtract 3 is the answer to the division example.

$$\begin{array}{rl} 12 & \\ \underline{-3} & (1) \\ 9 & \\ \underline{-3} & (2) \\ 6 & \\ \underline{-3} & (3) \\ 3 & \\ \underline{-3} & (4) \\ 0 & \end{array}$$

Since 3 was subtracted 4 times, you say that

$$3\overline{)12}^{\;4}$$

Division can be considered as (1) the inverse of multiplication or (2) repeated subtractions.

5.2 The Language Of Division

Example

What would be the share for each person if we divided $12 among 3 people?

Solution

To find out what each person would receive, you must divide $12 by 3. This could be written in any of the following ways:

$$3\overline{)12} \quad \text{or} \quad 12 \div 3 \quad \text{or} \quad \frac{12}{3} \quad \text{or} \quad 12/3$$

You see that there is more than one way to indicate division.

The various parts of the division example are:

$$\begin{array}{l} \quad\;\; 4 \leftarrow \text{quotient} \\ 3\overline{)12} \leftarrow \text{dividend} \\ \uparrow \text{——— divisor} \end{array}$$

We say that 3 divides *into* 12 or 12 is divided *by* 3. In either case, the *quotient* or answer is always 4.

You notice that division has a language too. There are terms you must know to understand division.

Terms You Should Remember

Divide To separate into groups or shares.

Dividend A number to be divided.

Divisor The number by which the dividend is to be divided.

Quotient The number resulting from the division of one number by another.

5.3 You Still Need Those Combinations

In each of the other three operations — addition, subtraction, multiplication — there was a total of 100 possible combinations. In division there are only 90. Do you know why?

It was stated earlier that division is the inverse operation of multiplication. This means that you find the quotient by multiplying the divisor by the quotient, which gives you a product equal to the dividend.

Look at the example $\frac{6}{0} = ?$, which is the same as $0\overline{)6}$.

This is a problem of finding the answer to the example $0 \times ? = 6$. Obviously there is *no* value which will make this a true statement. *Division by zero is meaningless, because there is no possible quotient.* Therefore, the divisor can be any number at all except 0.

This limits you to only 9 single-digit divisors and each one can be divided into 10 dividends. Thus, there are only 90 possible division combinations if we *limit* possible quotients to 1-digit numbers.

Dividing by 1 also has an interesting result. Look at these division examples.

$$1\overline{)5}^{\,?} \qquad 3 \div 1 = ? \qquad \frac{9}{1} = ?$$

Applying the inverse operation of multiplication for the first example, you find the quotient by multiplying the divisor (1) by the quotient (?) which gives you a product equal to the dividend (5), or $1 \times ? = 5$. The quotient is 5; $1 \times 5 = 5$. Similarly,

$$1 \times ? = 3 \qquad\qquad 1 \times ? = 9$$
$$? = 3 \qquad\qquad ? = 9$$

In general, dividing by 1 results in a quotient equal to the dividend.

What happens when you divide 0 by any number except 0?

$$\begin{array}{r} ? \\ 5\overline{)0} \end{array} \qquad 0 \div 9 = ? \qquad \frac{0}{1} = ?$$

Using the same procedure as before, you have

$$5 \times ? = 0 \qquad 9 \times ? = 0 \qquad 1 \times ? = 0$$

Each quotient is 0. In general, dividing 0 by any number except 0 results in a quotient of 0.

In the practice exercise which follows, all of the 90 possible division combinations are given. Review each of them carefully so you can gain speed and skill.

Again we suggest that you do *not* write the answers in the book. Write your answers on a separate piece of paper so you can use this exercise for practice as many times as you wish. The idea is to see how you can improve by recognizing the answers more and more quickly.

Remember that division relies on your power to multiply. Although you just completed the chapter on multiplication, it might be wise to return to chapter 4, practice exercises 41 and 42, and review the 100 multiplication combinations.

Practice Exercise 51

Divide.

1. $1\overline{)1}$ 2. $6\overline{)48}$ 3. $3\overline{)9}$ 4. $2\overline{)6}$ 5. $5\overline{)20}$

6. $8\overline{)16}$ 7. $4\overline{)4}$ 8. $6\overline{)42}$ 9. $5\overline{)0}$ 10. $9\overline{)9}$

11. $2\overline{)8}$ 12. $7\overline{)35}$ 13. $1\overline{)2}$ 14. $9\overline{)45}$ 15. $4\overline{)8}$

16. $4\overline{)0}$ 17. $7\overline{)42}$ 18. $3\overline{)27}$ 19. $9\overline{)54}$ 20. $5\overline{)10}$

21. $3\overline{)6}$ 22. $4\overline{)24}$ 23. $6\overline{)36}$ 24. $1\overline{)3}$ 25. $8\overline{)24}$

26. $2\overline{)14}$ 27. $7\overline{)7}$ 28. $9\overline{)72}$ 29. $5\overline{)40}$ 30. $3\overline{)0}$

31. $4\overline{)28}$ 32. $1\overline{)7}$ 33. $7\overline{)49}$ 34. $9\overline{)36}$ 35. $6\overline{)30}$

36. $9\overline{)0}$ 37. $5\overline{)35}$ 38. $3\overline{)3}$ 39. $9\overline{)63}$ 40. $8\overline{)32}$

41. $5\overline{)15}$ 42. $9\overline{)81}$ 43. $6\overline{)0}$ 44. $7\overline{)14}$ 45. $3\overline{)24}$

46. $8 \div 1 =$ 47. $32 \div 4 =$ 48. $48 \div 8 =$ 49. $24 \div 6 =$ 50. $10 \div 2 =$

51. $5\overline{)30}$ 52. $2\overline{)4}$ 53. $9\overline{)27}$ 54. $4\overline{)16}$ 55. $8\overline{)0}$

56. $3\overline{)21}$ 57. $6\overline{)18}$ 58. $7\overline{)56}$ 59. $1\overline{)5}$ 60. $5\overline{)5}$

61. $6\overline{)54}$ 62. $1\overline{)0}$ 63. $8\overline{)40}$ 64. $3\overline{)18}$ 65. $2\overline{)18}$

66. $\frac{12}{2} =$ 67. $\frac{9}{1} =$ 68. $\frac{12}{4} =$ 69. $\frac{18}{9} =$ 70. $\frac{25}{5} =$

71. $1\overline{)4}$ 72. $7\overline{)28}$ 73. $5\overline{)45}$ 74. $6\overline{)12}$ 75. $4\overline{)36}$

76. $7\overline{)0}$ 77. $3\overline{)12}$ 78. $8\overline{)72}$ 79. $7\overline{)63}$ 80. $2\overline{)2}$

81. $8\overline{)56}$ 82. $6\overline{)6}$ 83. $8\overline{)8}$ 84. $1\overline{)6}$ 85. $7\overline{)21}$

86. $2\overline{)16}$ 87. $4\overline{)20}$ 88. $8\overline{)64}$ 89. $2\overline{)0}$ 90. $3\overline{)15}$

5.4 Help From The Number Line

You represent division on the number line as a series of subtractions. Repeating the problem $12 \div 3$, you locate 12 on the number line and subtract groups of 3 units starting from the right.

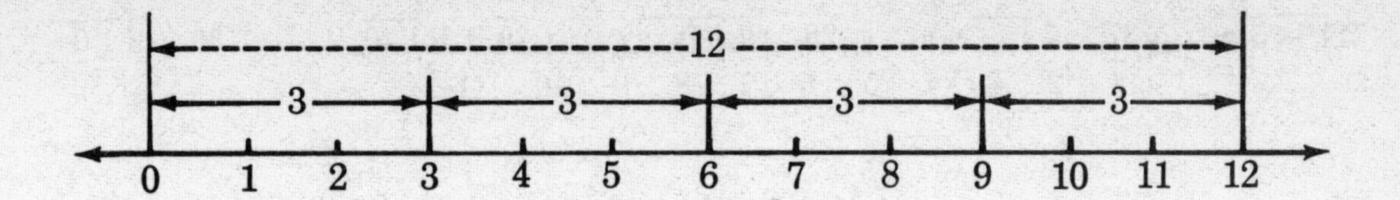

You subtract 4 equal groups of 3 units each, so

$$\frac{12}{3} = 4$$

Suppose you divide 13 by 5. How do you represent that division on the number line? Begin in the same manner as in the previous example. Subtract groups of 5 beginning from the right.

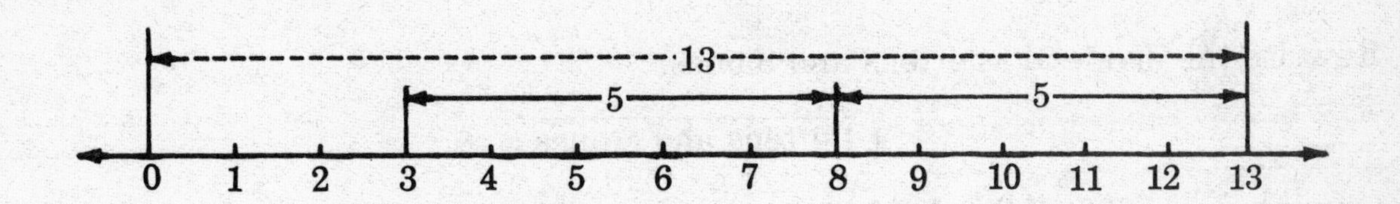

You see that you can only subtract *two* groups of 5. After subtracting two groups of 5, you end on 3 on the number line. You say that *two* groups of 5 may be subtracted and 3 units still remain; therefore 13 ÷ 5 = 2 with a remainder of 3. We write it as 13 ÷ 5 = 2 R 3 (R = remainder).

Practice Exercise 52

Illustrate each of the following division problems on a number line. Find each quotient.

1. 8 ÷ 2 =

2. $3\overline{)10}$

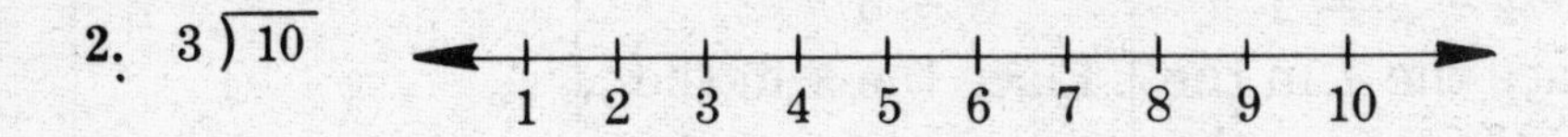

3. 16 ÷ 4 =

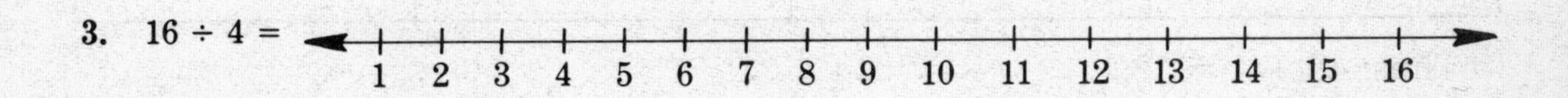

4. 11 ÷ 5 = 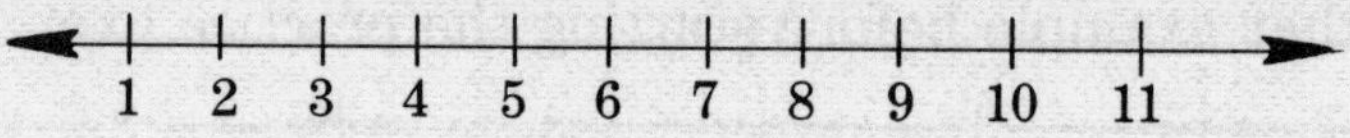

5.5 Dividing By A One-Digit Number

Example

A work crew of 48 men must be assigned to 4 different jobs. If each job requires an equal number of men, how many men will be assigned to each job?

Solution

To find out the size of each work crew, you must divide 48 by 4. It looks like this:

$$4\overline{)48}$$

Rewrite the dividend as 4 tens and 8 ones.

$$4\overline{)\text{4 tens and 8 ones}}$$

Dividing each part of the divident by 4 yields a quotient of

$$\begin{array}{l} \text{1 ten and 2 ones} \quad \text{or } 12 \\ 4\overline{)\text{4 tens and 8 ones}} \end{array}$$

Thus, you see that you can divide the divisor into the dividend in the following way.

Problem	Step 1	Step 2	Check
$4\overline{)48}$	$\begin{array}{r} 1 \\ 4\overline{)48} \end{array}$	$\begin{array}{r} 12 \\ 4\overline{)48} \end{array}$	$\begin{array}{r} 12 \\ \times\ 4 \\ \hline 48 \end{array}$
	Divide the divisor (4) into the *tens* place number (4): 4 ÷ 4 = 1 Place the 1 in the quotient over the 4 in the dividend.	Divide the divisor (4) into the *ones* place number (8): 8 ÷ 4 = 2 Place the 2 in the quotient over the 8 in the dividend.	

Thus: 48 ÷ 4 = 12.

Let's look at another example before starting the practice exercise.

Problem	Step 1	Step 2	Check
$3\overline{)63}$	$\begin{array}{r}2\\3\overline{)63}\end{array}$	$\begin{array}{r}21\\3\overline{)63}\end{array}$	$\begin{array}{r}21\\\times\ 3\\\hline 63\end{array}$
	Divide the divisor (3) into the *tens* number (6): $6 \div 3 = 2$ Place the 2 in the quotient over the *tens* number (6).	Divide the divisor (3) into the *ones* number (3): $3 \div 3 = 1$ Place the 1 in the quotient over the *ones* number (3).	

Thus: $63 \div 3 = 21$.

This same method is used regardless of the size of the number in the dividend. You begin by dividing into the number on the left. Then divide the next number and the next number until no more numbers remain.

Another method of doing a problem in division is by the process of *estimation*. Let's return to the problem:

$$4\overline{)48}$$

You estimate the number of times 4 will divide into 48. Let's guess the number is 10. Find the product of the divisor (4) and your estimate (10): $4 \times 10 = 40$. Since 10 did not make the product come out evenly, 10 is only a *partial quotient*. Subtracting the 40 from the dividend of 48 leaves you with a remainder of 8. To divide the remainder (8) by the divisor (4) we do not need to estimate, since we know that $8 \div 4 = 2$. This is a second *partial quotient* which we add to the first one. Our full quotient is: $10 + 2 = 12$.

Problem	Step 1	Step 2	Step 3
$4\overline{)48}$	$\begin{array}{r}10\\4\overline{)48}\end{array}$	$\begin{array}{r}10\\4\overline{)48}\\\underline{40}\\8\end{array}$	$\begin{array}{r}2\\10\\4\overline{)48}\\\underline{40}\\8\\\underline{8}\end{array}$ $\begin{array}{r}2\\\underline{+10}\\12\end{array}$
	Estimate the number of times 4 will divide into 48. Any estimate is good so long as the product of the divisor and estimate is not higher than 48.	Multiply the divisor by your estimate: $4 \times 10 = 40$ Your partial quotient is 10. Subtract the product from the dividend: $48 - 40 = 8$	Divide the divisor (4) into the remainder (8): $8 \div 4 = 2$ Add the 2 to the first partial quotient (10): $10 + 2 = 12$

Thus: $48 \div 4 = 12$.

Let's look at another example using this same method of estimation.

Problem	Step 1	Step 2	Step 3
$3\overline{)204}$	$\begin{array}{r}60\\3\overline{)204}\end{array}$	$\begin{array}{r}60\\3\overline{)204}\\\underline{180}\\24\end{array}$	$\begin{array}{r}8\\60\\3\overline{)204}\\\underline{180}\\24\\\underline{24}\end{array}$ $\begin{array}{r}8\\\underline{+60}\\68\end{array}$
	Estimate the number of times 3 will divide into 204, for example, 60. The first partial quotient is 60.	Multiply: $3 \times 60 = 180$ Subtract: $204 - 180 = 24$	Divide the divisor (3) into the remainder (24): $24 \div 3 = 8$ Add the 8 to the first partial quotient (60): $60 + 8 = 68$

Thus: $204 \div 3 = 68$.

And another —

Problem	Step 1	Step 2	Step 3	Step 4
9) 2,865	300 9) 2,865	300 9) 2,865 2,700 165	10 300 9) 2,865 2,700 165 90 75	8 10 300 9) 2,865 2,700 165 90 75 72 8 10 +300 318 R 3
	Estimate: 9) 2865 One estimate is 300.	Multiply: 9 × 300 = 2700 300 is the first partial quotient. Subtract: 2865 − 2700 = 165	Estimate: 9) 165 One estimate is 10. Multiply: 9 × 10 = 90 Subtract: 165 − 90 = 75 The second partial quotient is 10.	Divide: 75 ÷ 9 = 8 R/3 The third partial quotient is 8. Add: 300 + 10 + 8 = 318

The complete quotient is 9) 2,865 318 R 3.

Practice Exercise 53

Divide. Check by multiplying.

1. 3) 93
2. 4) 84
3. 5) 55
4. 2) 86
5. 3) 69
6. 4) 48
7. 8) 88
8. 2) 62
9. 6) 66
10. 3) 36

Example

There are 488 employees in the A & R Auto Company. How can they be divided so that an equal number will be sent to 4 different departments?

Solution

Divide as shown below.

Problem	Step 1	Step 2	Step 3	Check
$4\overline{)488}$	$\begin{array}{r}1\\4\overline{)488}\end{array}$	$\begin{array}{r}12\\4\overline{)488}\end{array}$	$\begin{array}{r}122\\4\overline{)488}\end{array}$	$\begin{array}{r}122\\\times\ \ 4\\\hline 488\end{array}$
	Divide: $4 \div 4 = 1$ Place the 1 in the quotient over the *hundreds* number (4).	Divide: $8 \div 4 = 2$ Place the 2 in the quotient over the *tens* number (8).	Divide: $8 \div 4 = 2$ Place the 2 in the quotient over the *ones* number (8).	

Thus: $488 \div 4 = 122$.

Example

Three partners earn a total of $36,096 yearly. How much does each one earn?

Solution

Look at this problem:

Problem	Step 1	Step 2	Step 3
$3\overline{)\$36{,}096}$	$\begin{array}{r}1\\3\overline{)\$36{,}096}\end{array}$	$\begin{array}{r}12\\3\overline{)\$36{,}096}\end{array}$	$\begin{array}{r}12{,}0\\3\overline{)\$36{,}096}\end{array}$
	Divide: $3 \div 3 = 1$	Divide: $6 \div 3 = 2$	Divide: $0 \div 3 = 0$

Step 4	Step 5	Check
$\begin{array}{r}12{,}03\\3\overline{)\$36{,}096}\end{array}$	$\begin{array}{r}\$12{,}032\\3\overline{)\$36{,}096}\end{array}$	$\begin{array}{r}\$12{,}032\\\times\qquad 3\\\hline \$36{,}096\end{array}$
Divide: $9 \div 3 = 3$	Divide: $6 \div 3 = 2$	

Thus: $\$36{,}096 \div 3 = \$12{,}032$.

Practice Exercise 54

Divide. Check each answer by multiplying.

1. $3\overline{)669}$ 2. $4\overline{)804}$ 3. $2\overline{)8{,}642}$ 4. $5\overline{)505}$ 5. $3\overline{)9{,}336}$

6. $6\overline{)606}$ 7. $2\overline{)842}$ 8. $4\overline{)8{,}044}$ 9. $3\overline{)39{,}303}$ 10. $8\overline{)8{,}888}$

11. $2\overline{)20{,}024}$ 12. $6\overline{)666}$ 13. $3\overline{)3{,}036}$ 14. $3\overline{)966}$ 15. $2\overline{)868}$

5.6 When The Quotient Contains Fewer Places Than The Dividend

Have you noticed in each example you have done thus far that the quotient contains as many places as the dividend? That happens when the divisor divides into the first number on the left. What happens if it does not?

Example

Five cartons weigh 455 lb. Each carton weighs the same. How many pounds does each carton weigh?

Solution

Dividing 455 by 5 will give the answer to the problem.

Problem	Step 1	Step 2	Check
$5\overline{)455}$	$\begin{array}{r}9\\5\overline{)455}\end{array}$	$\begin{array}{r}91\\5\overline{)455}\end{array}$	$\begin{array}{r}91\\\times\ 5\\\hline 455\end{array}$
	Divide the 5 into the *hundreds* number (4): $4 \div 5 = ?$ *or* $5 \times ? = 4$ There is no whole number which fits, so divide the 5 into the first two numbers (45): $45 \div 5 = 9$ Place the 9 in the quotient over the *tens* number (5).	Divide the 5 into the *ones* number (5): $5 \div 5 = 1$ Place the 1 in the quotient over the *ones* number (5).	

Thus: $455 \div 5 = 91$.

Practice Exercise 55

Divide. Check by multiplying.

1. $9\overline{)549}$ 2. $8\overline{)248}$ 3. $3\overline{)216}$ 4. $7\overline{)357}$ 5. $9\overline{)729}$

6. $4\overline{)324}$ 7. $5\overline{)155}$ 8. $7\overline{)217}$ 9. $2\overline{)128}$ 10. $6\overline{)486}$

5.7 Zero As A Place Holder

Example

Alba earns $312 for 3 weeks' work. How much does she earn each week?

Solution

The quotient of 312 ÷ 3 will tell us what Alba earns weekly. Sometimes you will find that it is not the first number on the left in the dividend which is too small to divide but another number in the dividend. In this problem, it is the 1. Look at the following illustration:

Problem	Step 1	Step 2	Step 3	Check
3) $312	1 3) $312	10 3) $312	$104 3) $312	$104 × 3 $312
	Divide: 3 ÷ 3 = 1 Place the 1 in the quotient above the 3.	Divide: 1 ÷ 3 = ? *or* 3 × ? = 1 This is not possible, so *place a zero in the quotient above the tens number (1).*	Divide 3 into the *tens* and *ones* numbers (12): 12 ÷ 3 = 4 Place the 4 above the *ones* number in the quotient.	

Thus: $312 ÷ 3 = $104.

Example

A stereo set costs $220. You put $60 down and agree to pay the remainder in 8 equal monthly installments. How much must you pay each month?

Solution

Since the set costs $220 and you put $60 down, the balance you owe is only $220 – $60 = $160. If you are going to pay $160 in 8 months, you must divide to find your monthly payments. Let's look at this problem:

Problem	Step 1	Step 2	Check
8) $160	2 8) $160	$ 20* 8) $160	$20 × 8 $160
	Divide: 1 ÷ 8 = ? *or* 8 × ? = 1 This is not possible, so divide the first two numbers (16) by 8: 16 ÷ 8 = 2 Place the 2 in the quotient above the *tens* number (6).	Divide: 0 ÷ 8 = 0 Place the answer (0) in the quotient over the *ones* number (0).	

*Don't forget to include the *zero* in the quotient. Omitting the zero gives a quotient of 2, which does not check when multiplying.

Thus: $160 ÷ 8 = $20.

Practice Exercise 56

Divide. Check each answer by multiplying. Don't forget to include the zero when it is needed.

1. 6) 624 **2.** 7) 280 **3.** 5) 525 **4.** 3) 318

5. 7) 735 **6.** 8) 720 **7.** 9) 810 **8.** 8) 816

9. 2) 2,610 **10.** 4) 4,124 **11.** 3) 3,660 **12.** 2) 2,120

5.8 Dividing And Leaving A Remainder

In Section 5.4, "Help From the Number Line," an example

$$5\overline{)13}$$

was illustrated. It was found that two groups of 5 could be subtracted and there would be 3 units remaining. Therefore, we say that $13 \div 5 = 2$ with a *remainder* of 3. A short way of writing the remainder is R, so the problem would look like this:

Problem	Step 1	Step 2	Check
$5\overline{)13}$	$\begin{array}{r}2\\5\overline{)13}\end{array}$	$\begin{array}{r}2\text{ R }3\\5\overline{)13}\end{array}$	$\begin{array}{r}5\\\times2\\\hline10\end{array}$ $\begin{array}{r}10\\+\ 3\\\hline13\end{array}$
	Since 5 does not divide into 1 you divide it into the two numbers (13): $13 \div 5 = ?$ *or* $5 \times ? = 13$. *No number divides this evenly*, for 2 is too small and 3 is too large: $5 \times 2 = 10$ and $5 \times 3 = 15$. The smaller one (2) is our answer so place it in the quotient above the *ones* number (3).	$5 \times 2 = 10$ This leaves a remainder of 3.	

Thus: $\begin{array}{r}2\text{ R }3.\\5\overline{)13}\end{array}$

In the check you multiply the quotient (2) by the divisor (5) and add the remainder (3) to get the dividend (13).

$$\begin{array}{r}2 \times 5 = 10\\+\ \ 3\\\hline13\end{array}$$

Practice Exercise 57

Divide. Check each answer by multiplying.

1. $2\overline{)17}$ 2. $5\overline{)39}$ 3. $7\overline{)50}$ 4. $9\overline{)46}$ 5. $4\overline{)19}$

6. $3\overline{)25}$ 7. $8\overline{)35}$ 8. $3\overline{)20}$ 9. $6\overline{)47}$ 10. $9\overline{)83}$

11. $6\overline{)45}$ 12. $4\overline{)37}$ 13. $8\overline{)67}$ 14. $5\overline{)17}$ 15. $7\overline{)26}$

5.9 Carrying Remainders

Example

Sixty-five people volunteer for 5 different jobs. If an equal number are assigned to each job, how many people will be on each job?

Solution

Once again the problem calls for dividing one number by another. The problem is:

$$5\overline{)65}$$

Rewriting the example, as done previously, looks like this:

$$5\overline{)\text{6 tens and 5 ones}}$$

Dividing 6 tens and 5 ones into *five* groups is difficult but since 5 divides evenly into 5 tens, let's rewrite the dividend as 5 tens + 1 ten + 5 ones, or 5 tens and 15 ones (recall that 1 ten = 10 ones).

$$5\overline{)\text{5 tens and 15 ones}}$$

Dividing each part of the dividend by 5 yields a quotient of

$$\begin{array}{r} \text{1 ten and 3 ones} \quad \text{or} \quad 13 \\ 5\overline{)\text{5 tens and 15 ones}} \end{array}$$

Following the same procedure in the original example would look like this:

Problem	Step 1	Step 2	Check
$5\overline{)65}$	$\begin{array}{r} 1 \\ 5\overline{)65} \end{array}$	$\begin{array}{r} 13 \\ 5\overline{)65} \end{array}$	$\begin{array}{r} 13 \\ \times\ 5 \\ \hline 65 \end{array}$
	Divide the divisor (5) into the *tens* number (6): 6 ÷ 5 = ? cannot be divided evenly. The answer is 1 with a remainder of 1. The answer (1) is placed in the quotient above the *tens* number (6). The remainder (1) or one ten is carried to the *ones* number (5), making it 15.	Divide the 5 into the new *ones* number (15): 15 ÷ 5 = 3. Place the 3 in the quotient above the *ones* number (5).	

Thus: 65 ÷ 5 = 13.

This same method can be applied to any example where the divisor doesn't divide the first number or any other number evenly. Look at the next two examples before attempting the practice exercise that follows.

Example

The sum of $519 is to be divided evenly among 3 people. How much money does each one receive?

Solution

Problem	Step 1	Step 2	Step 3	Check
$3\overline{)\$519}$	$\begin{array}{r}1\\3\overline{)\$519}\end{array}$	$\begin{array}{r}17\\3\overline{)\$519}\end{array}$	$\begin{array}{r}\$173\\3\overline{)\$519}\end{array}$	$\begin{array}{r}\$173\\\times\ \ 3\\\hline\$519\end{array}$
	Divide: $5 \div 3 = 1$ R 2 Place the 1 in the quotient above the 5 and carry the remainder 2 to the *tens* number. The *tens* number is now 21.	Divide: $21 \div 3 = 7$ Place the 7 above the *tens* number (1).	Divide: $9 \div 3 = 3$ Place the 3 above the *ones* number (9).	

Thus: $\$519 \div 3 = \$173$.

Example

Six clerks handle accounts for 768 customers. If each clerk handles an equal share, how many customers does each one have?

Solution

Problem	Step 1	Step 2	Step 3	Check
$6\overline{)768}$	$\begin{array}{r}1\\6\overline{)768}\end{array}$	$\begin{array}{r}12\\6\overline{)768}\end{array}$	$\begin{array}{r}128\\6\overline{)768}\end{array}$	$\begin{array}{r}128\\\times\ \ 6\\\hline768\end{array}$
	Divide: $7 \div 6 = 1$ R 1 Place the 1 in the quotient and carry 1, making the *tens* place 16. *Remember*— 1 hundred = 10 tens.	Divide: $16 \div 6 = 2$ R 4 Place the 2 in the quotient and carry the 4, making the *ones* place 48. *Remember*— 4 tens = 40 ones.	Divide: $48 \div 6 = 8$ Place the 8 in the quotient.	

Thus: $768 \div 6 = 128$.

Practice Exercise 58

Divide. Check each answer by multiplying. Be sure to indicate the remainder if there is one.

1. $6\overline{)78}$
2. $3\overline{)249}$
3. $4\overline{)76}$
4. $7\overline{)497}$
5. $3\overline{)231}$
6. $5\overline{)146}$
7. $2\overline{)1,302}$
8. $6\overline{)4,746}$
9. $2\overline{)1,359}$
10. $8\overline{)6,504}$
11. $9\overline{)7,002}$
12. $7\overline{)23,817}$

5.10 Word Problems Requiring Division

Reading word problems and understanding their solution require skills that have been presented already. The need is for more practice so that these habits come easily. Key words associated with solving problems by division have been enumerated in the illustrated examples already. The key word most commonly found in each problem is *each*. "How much does *each* one receive?"

Now try these problems.

Practice Exercise 59

1. Your daughter brings home 5 friends for cookies and milk. If there are 12 cookies to be divided equally among the 6 children, how many cookies does *each* one receive?

2. Tomato juice is on sale at 3 cans for 57¢. How many cents does *one* can cost?

3. A boat makes 7 trips and returns with 350 lb. of scallops. It brings back an equal amount for each trip. How many pounds does it bring back *each* time?

4. Mr. Gandolfo made 144 gal. of wine and must place it in barrels that *each* hold 9 gal. How many barrels does he need?

5. Mike makes 4 telephone calls and is charged 52¢. How much does *each* call cost?

6. The annual prize money of $210 is to be divided equally among the first 3 winners. How much money will *each* of the winners receive?

7. You put $400 down on an automobile which costs $2,400 and agree to pay the rest in 8 equal installments. How much is *each* monthly payment?

8. A truck is carrying 6 equal bales of cotton whose total weight is 2,070 lb. What is the weight of *each* bale?

9. A salesman drives on the average 424 mi. each week. If his car uses 1 gal. of gasoline for each 8 mi. he drives, how many gallons of gasoline does he use *each* week?

10. Bill finds that he has spent $312 in the past 6 months for the purchase of phonograph records. How many dollars has he spent *each* month if he spends an equal amount each month?

5.11 Dividing By More Than A One-Digit Number

When dividing by more than a one-digit number, the process of estimation is best used. The method is identical to the one described earlier. Since you will be multiplying the divisor by numbers like 10, 20, 50, 100, etc., it is advisable for you to review Practice Exercise 46 in Chapter 4, before you continue. Be sure you remember the short-cut method for multiplying before you look at the example.

Example

If 672 parking tickets were issued in a 32-day period, approximately how many were issued each day?

Solution

Dividing 672 by 32 will give the answer to this problem. Use the procedure that follows:

Problem	Step 1	Step 2	Step 3	Check
32) 672	20 32) 672	20 32) 672 640 32	1 20 20 + 1 32) 672 21 640 32 32 00	32 ×21 32 64 672
	To find the number of times 32 will divide into 672, you estimate the number of 30s in 600. This is 20. Place this first partial quotient in your answer.	Multiply: 32 × 20 = 640 Subtract this product from the dividend 672. 672 − 640 = 32	Divide: 32 ÷ 32 = 1 The second partial quotient of 1 is added to the 20: 20 + 1 = 21	

Thus: 672 ÷ 32 = 21.

You see that division by any number of digits is treated the same as before. To be sure you understand more difficult examples, two more illustrations will follow.

Example

Find the quotient of 5,544 divided by 42.

Solution

Problem	step 1	Step 2	Step 3	Check
42) 5,544	100 42) 5,544 4,200 1,344	30 100 42) 5,544 4,200 1,344 1,260 84	2 2 30 30 100 +100 42) 5,544 132 4,200 1,344 1,260 84 84	132 × 42 264 5 28 5,544
	Estimate as 100 the number of times 42 divides into 5544, Multiply: 42 × 100 = 4200 *Be sure this product is either less than or equal to the dividend.* Subtract: 5544 − 4200 = 1344	Estimate as 30 the number of times 42 divides into 1344: 42 × 30 = 1260 Subtract: 1344 − 1260 = 84	Divide: 84 ÷ 42 = 2 Add all the estimates or partial quotients together to find the final quotient: 100 + 30 + 2 = 132	

Thus: 5,544 ÷ 42 = 132.

You see that the partial quotients are successively 100, 30, and 2, thus giving 132 as the quotient. Since the first partial quotient is 100, you need only to write 1 in the hundreds place. Since the second partial quotient is 30, you need only to write 3 in the tens place. Since the third partial quotient is 2, you need only to write 2 in the ones place.

Thus you can see that the problem 5,544 ÷ 42 = ? can be written in a short form:

```
       132
42 ) 5,544
     4,200
     1,344
     1,260
        84
        84
```

Divisions may also have remainders as illustrated in this example.

Example

What is the quotient and remainder when 6,578 is divided by 365?

Solution

Problem	Step 1	Step 2	Step 3	Check
365) 6,578	2 365) 6,578 7,300	1 365) 6,578 3,650 2,928	18 = 18 365) 6,578 R 8 3,650 2,928 2,920 8	365 × 18 2,920 3 65 6,570 + 8 6,578
	Estimate: 6578 ÷ 365 = 20 Multiply: 365 × 20 = 7300 This product is larger than the dividend, 6578, so your estimate is too large.	Re-estimate: the partial quotient as 10. Multiply: 365 × 10 = 3650 Place the 1 in the quotient above the *tens* number (7). Subtract: 6578 − 3650 = 2928	Estimate: 2928 ÷ 365 = 8 Multiply: 365 × 8 = 2920 Place the 8 in the quotient above the *ones* number (8). Subtract: 2928 − 2920 = 8	

Thus: 6,578 ÷ 365 = 18 R 8.

A remainder must always be less than the divisor. It should be divided again if it is not.

Practice Exercise 60

Divide. Check each answer by multiplying. Some of the exercises have remainders. Try using the short-form method described earlier.

1. $13\overline{)78}$	2. $22\overline{)286}$	3. $12\overline{)468}$	4. $23\overline{)1,058}$
5. $34\overline{)102}$	6. $18\overline{)60}$	7. $52\overline{)1,352}$	8. $15\overline{)17,130}$
9. $28\overline{)2,548}$	10. $94\overline{)8,272}$	11. $45\overline{)4,590}$	12. $68\overline{)4,914}$
13. $74\overline{)1,336}$	14. $123\overline{)7,011}$	15. $77\overline{)9,920}$	16. $278\overline{)834}$

Terms You Should Remember

Remainder The dividend minus the product of the divisor and quotient.

Estimate To calculate approximately.

Partial quotient An incomplete quantity resulting from dividing part of the quotient by the divisor.

Review Of Important Ideas

Some of the most important ideas thus far in Chapter 5 are:

Division is repeated subtraction.

Division is the inverse operation of multiplication.

Estimation is one method for doing division examples.

Division depends on your ability to multiply.

Remainders may occur when dividing. All problems do not divide evenly.

5.12 More Difficult Word Problems Requiring Division

Read section 5.10 to recall the key words associated with problems solved by dividing before doing this next exercise.

Practice Exercise 61

1. There are 216 adults attending 8 classes of English in the evening adult high school. On the average, how many were enrolled in each class?

2. You earn \$7,476 in a year. How many dollars do you earn monthly? (12 months = 1 year)

3. You purchase 2 snowmobiles for \$1,128 and agree to pay for them in 24 equal monthly payments. What is your monthly payment?

4. A truck is carrying a load of 108 barrels. The total weight of the barrels is 6,804 lb. If each barrel is the same weight, what does one barrel weigh?

5. A man purchases a piece of lumber 192 in. long. How many pieces 16 in. long can be cut from it? (Disregard any waste occurring from cutting.)

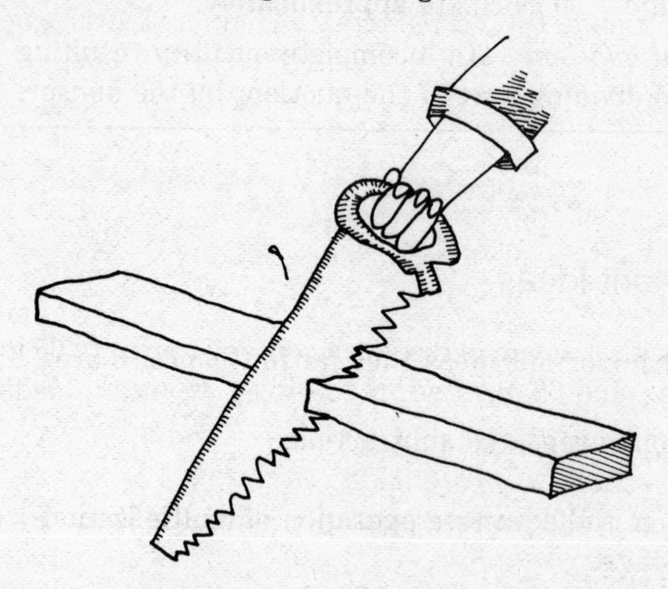

6. Mr. Jones, at the end of 47 weeks of work, had earned \$7,755. How many dollars did he earn weekly?

7. A car rental agency rents 16,464 cars each year. On the average, how many cars do they rent monthly? (12 months = 1 year)

8. A real estate salesman travels 420 mi. each week. If his car uses 1 gal. of gasoline for each 12 mi. he drives, how many gal. of gasoline does he use in each week?

9. June requires 54 in. of material to make a skirt. If she orders 432 in. of cloth, how many skirts can she make?

10. A truck carries bales of cotton whose total weight is 13,875 lb. If each bale weighs 375 lb., how many bales are on the truck?

5.13 Average

There are *three* different methods for determining the average. The first and most common is the *mean*. The other two are the *median* and the *mode*. All three are referred to as measures of central tendency. This means that each of them is a value near the middle or center of all the other numbers when they are arranged in order of size. The method for obtaining each one suggests its definition.

1. *Mean* — Find the mean by adding the given values and then dividing the sum by the number of addends.

2. *Median* — Arrange the values in ascending or descending order and find that value which is in the center of all the numbers.

3. *Mode* — Find the value among all the numbers which occurs most frequently. There may or may not be a mode.

In order for you to get a better idea of how these *three* measures of central tendency behave, look at the following example:

Example

If the heights of the first seven persons who enter a store are 65 in., 72 in., 75 in., 67 in., 60 in., 65 in., and 65 in., find, if possible, the mean, median, and mode.

Solution

The listed heights of the seven persons are called the *data*. From the data we determine our averages.

1. *Mean* — 65 in. + 72 in. + 75 in. + 67 in. + 60 in. + 65 in. + 65 in. = 469 in. There are 7 addends, so 469 ÷ 7 = 67 in.
2. *Median* — 60 in., 65 in., 65 in., (65 in.), 67 in., 72 in., 75 in. The center value is the fourth one, 65 in.
3. *Mode* — The value of 65 in. occurs three times while all the others occur once. Thus, 65 in. is the mode.

Thus: The mean height is 67 in.
The median height is 65 in.
The mode height is 65 in.

The mean, median, and mode may be the same or may be different, as just shown. You do see, through, that they all tend toward the center of the data.

Practice Exercise 62

1. The four members of the Russo family weigh 132 lbs., 117 lbs., 94 lbs., and 73 lbs. What is their mean weight?

2. Charles received the following grades on his math tests: 85%, 94%, 64%, 88%, and 89%.

   a. What was the mean grade? ____________________

   b. What was the median grade? ____________________

   c. Is there a mode? ____________ If so, what is it? ____________________

3. In one week of driving his car, Mr. Betancourt used 13, 11, 5, 11, 10, 9, and 4 gal. of gasoline each day.

   a. What was the mean number of gallons used per day? ____________________

   b. What was the median number of gallons used per day? ____________________

   c. Is there a mode? ____________ If so, what is it? ____________________

4. If the total weight of the 11 members of a football team was 2,552 lb., then what is the average weight (mean) for each member of the team?

5. If the Cartwright family of 5 took a trip and spent a total of $365, then what was the average expense (mean) for each member of the family?

6. Mary, an office worker, timed herself at various times of the day to determine the average (mean) number of words typed in one minute. She found that as the day progressed she typed with less speed. These were her findings: 63, 61, 57, 54, 43, and 40 words per minute. What was her average number of words per minute?

5.14 Division Applied To Area Measurement

Example

Mr. Sims had a gallon of paint which could cover 520 sq. ft. of rectangular wall surface. If the height of the wall is 8 ft., then what is the length, in feet, of the wall surface that can be painted with his gallon of paint?

Solution

Since the walls are rectangular in shape, the problem becomes simply an example in which you know the area of the rectangle and its width. You are looking for its length.

$$8 \text{ ft.} \times ? = 520 \text{ sq. ft.}$$
$$? = 65 \text{ ft.}$$

Area = 520 sq. ft. x'

8′

Thus, Mr. Sims can cover 65 ft. of wall with his gallon of paint.

Practice Exercise 63

1. A rectangular garden contains 1,462 sq. ft. of surface that can be planted. If the length of the rectangle is 43 ft., find its width.

2. If the area of a rectangular plot of ground contains 612 sq. ft. and its length is 34 ft., what is its width?

3. The printed material on a rectangular sheet of paper covered 108 sq. in. If its length is 12 in., find its width.

Check What You Have Learned

The following test lets you see how well you have learned the ideas in Chapter 5.

Posttest 5

Write each answer in the space provided.

1. $7\overline{)35}$
2. $84 \div 4 =$
3. $2\overline{)628}$
4. $6\overline{)606}$
5. $3\overline{)186}$
6. $9\overline{)990}$
7. $\frac{51}{9} =$
8. $348 \div 2 =$
9. $3\overline{)618}$
10. $4\overline{)512}$

11. What would be the share for each person if we divide \$16 among 8 people?

12. Illustrate $9 \div 3$ on the number line below.

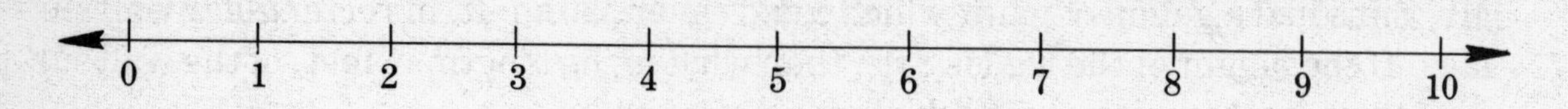

13. Six people buy an equal share in the purchase of a lottery ticket. If they win \$54,048, how much does each one receive?

14. $76 \div 19 =$

15. $26\overline{)546}$

16. $42\overline{)2{,}184}$

17. $32\overline{)4{,}626}$

18. Find the mean, median, and mode for the following data: 32 ft., 30 ft., 41 ft., 38 ft., 36 ft., 30 ft., 32 ft., 32 ft., 35 ft.

19. The area of a rectangle contains 138 sq. yd. If the length is 23 yd., find the width.

A= 138 sq. yds.

? yds

23 yds

20. At a recent fund-raising dinner, \$78,500 was collected from the sale of \$25 tickets. How many people attended the affair?

ANSWERS AND EXPLANATIONS TO POSTTEST 5

1. $\begin{array}{r}5\\7\overline{)35}\end{array}$

2. $\begin{array}{r}21\\4\overline{)84}\end{array}$

3. $\begin{array}{r}314\\2\overline{)628}\end{array}$

4. $\begin{array}{r}101\\6\overline{)606}\end{array}$

5. $\begin{array}{r}62\\3\overline{)186}\end{array}$

6. $\begin{array}{r}110\\9\overline{)990}\end{array}$

7. $\frac{51}{9} = 5 \text{ R } 6$

8. $\begin{array}{r}174\\2\overline{)348}\end{array}$

9. $\begin{array}{r}206\\3\overline{)618}\end{array}$

10. $\begin{array}{r}128\\4\overline{)512}\end{array}$

11. $\$16 \div 8 = \$2$

12. $9 \div 3 = 3$

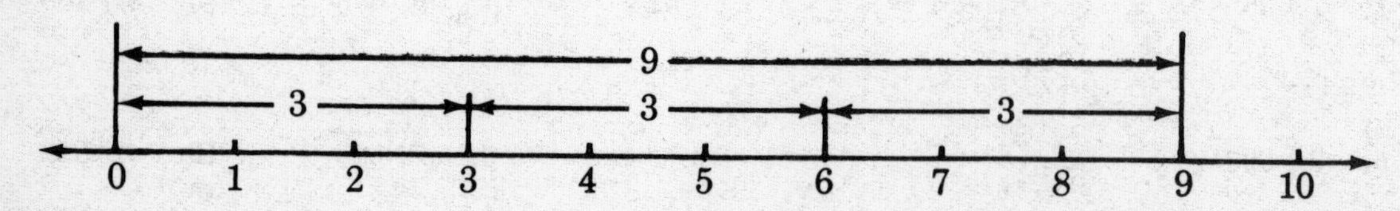

13. $6 \overline{) \$54,048}$ = \$ 9,008

14. $19 \overline{) 76}$ = 4

15. $26 \overline{) 546}$ = 21
```
     21
26 ) 546
     520
     ---
      26
      26
```

16.
```
       52
42 ) 2,184
     2,100
     -----
        54
        54
```

17.
```
      144 R 18
32 ) 4,626
     3,200
     -----
     1,426
     1,280
     -----
       146
       128
       ---
        18
```

18. 30 ft., 30 ft., 32 ft., 32 ft., 32 ft., 35 ft., 36 ft., 38 ft., 41 ft.

$$\text{mean} = \frac{30+30+32+32+32+35+36+38+41}{9} = \frac{306}{9} = 34 \text{ ft.}$$

median = 30, 30, 32, 32, (32), 35, 36, 38, 41 = 32 ft.

mode = 32 ft.

19. $23 \times ? = 138$
$? = 6$ yds.

20.
```
        $ 3,140
$25 ) $78,500
       75,000
       ------
        3,500
        2,500
        -----
        1,000
        1,000
```

In counting up your answers, remember that there were 22 separate answers.

A Score of	*Means That You*
20–22	Did very well. You can move on to Chapter 6 in Volume 2.
17–19	Know the material except for a few points. Reread the sections about the ones you missed.
14–16	Need to check carefully the sections you missed.
0–13	Need to review the chapter again to refresh your memory and improve your skills.

Questions	*Are Covered in Section*
1	5.3
12	5.4
2–4	5.5
5	5.6
6, 9	5.7
7	5.8
8, 10	5.9
11, 13	5.10
14–17	5.11
20	5.12
18	5.13
19	5.14

Division of Whole Numbers

ACROSS

1. $6\overline{)72}$
3. $3\overline{)48}$
4. $7\overline{)1,981}$
5. $5\overline{)125}$
6. $6\overline{)360}$
7. $12\overline{)192}$
10. $5\overline{)31,125}$
12. $24\overline{)89,304}$
15. $73\overline{)730}$
16. $324 \div 18 =$
18. $216 \div 6 =$
19. $4\overline{)3,276}$
20. $11\overline{)781}$
21. $1,040 \div 13 =$

DOWN

1. $6\overline{)1,086}$
2. $4\overline{)92}$
3. $17\overline{)255}$
5. $6\overline{)1,236}$
7. $23\overline{)276}$
8. $650 \div 10 =$
9. $2,736 \div 38 =$
11. $9\overline{)261}$
12. $17\overline{)527}$
13. $910 \div 13 =$
14. $11\overline{)1,276}$
17. $15\overline{)6,150}$
18. $186 \div 6 =$
19. $528 \div 6 =$

You will find the solution on page 167.

Puzzle Time

All Operations

ACROSS

1. 86 − 49 =
3. 8 × 7 =
4. 686 ÷ 7 =
6. 9 × 7 =
8. 7 × 2 =
9. 10 × 9 + 2 =
11. 16 × 6 =
12. 2 × 7 + 1 =
13. 12 × 11 =
14. 2,744 ÷ 2 =
16. 11 × 11 =
18. 11 × 8 − 10 =
19. 6 × 97 − 564 =
21. 93 × 3 + 5 =
23. 111 × 7 =
26. 2,852 ÷ 4 =
27. 30 × 7 =
29. 7 × 7 − 7 =
31. 10 × 5 =
33. 1,776 ÷ 3 =
35. 8,060 − 939 =
38. 444 − 321 =
41. 9 × 8 − 23 =
43. 4 × 8 =
44. 7 × 6 + 1 =
45. 14 × 4 =
47. 12 × 5 =
48. 10 × 9 =
49. 13 × 3 =
50. 12 × 8 − 11 =

DOWN

1. $3 \times 12 - 5 =$
2. $9{,}583 - 2{,}100 =$
3. $181 \times 3 =$
5. $4{,}060 \div 5 =$
6. $2{,}304 \times 3 =$
7. $6 \times 6 =$
9. $2{,}739 \div 3 =$
10. $512 - 255 =$
13. $7 \times 3 - 2 =$
14. $9 \times 2 =$
15. $7 \times 3 =$
17. $156 \div 13 =$
18. $44{,}604 \div 6 =$
20. $1{,}744 \times 5 =$
21. $9 \times 3 =$
22. $9 \times 9 =$
24. $14 \times 4 + 15 =$
25. $10 \times 7 =$
28. $7 \times 5 =$
30. $9 \times 3 =$
31. $204 \div 4 =$
32. $7 \times 9 =$
34. $12{,}893 - 3{,}573 =$
36. $12 \times 12 =$
37. $63 \times 4 + 41 =$
39. $7{,}374 \div 3 =$
40. $13 \times 13 =$
42. $146 \times 3 + 1 =$
43. $6 \times 6 =$
46. $14 \times 5 - 5 =$

The solution to this puzzle is on page 167.

ANSWERS FOR CHAPTER 5

PRETEST 5

1. 7
2. 23
3. 213
4. 101
5. 52
6. 110
7. 7 R 2
8. 171
9. 206
10. 128
11. $7
12. $6 \div 2 = 3$

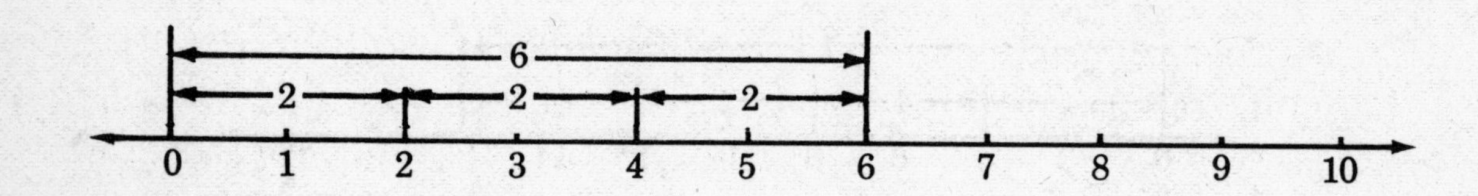

13. $8,007
14. 5
15. 21
16. 52
17. 144 R 18
18. mean = 63 in.
median = 62 in.
mode = 62 in.
19. 6 ft.
20. $245

PRACTICE EXERCISE 51

1. 1
2. 8
3. 3
4. 3
5. 4
6. 2
7. 1
8. 7
9. 0
10. 1
11. 4
12. 5
13. 2
14. 5
15. 2

16. 0	**17.** 6	**18.** 9	**19.** 6	**20.** 2
21. 2	**22.** 6	**23.** 6	**24.** 3	**25.** 3
26. 7	**27.** 1	**28.** 8	**29.** 8	**30.** 0
31. 7	**32.** 7	**33.** 7	**34.** 4	**35.** 5
36. 0	**37.** 7	**38.** 1	**39.** 7	**40.** 4
41. 3	**42.** 9	**43.** 0	**44.** 2	**45.** 8
46. 8	**47.** 8	**48.** 6	**49.** 4	**50.** 5
51. 6	**52.** 2	**53.** 3	**54.** 4	**55.** 0
56. 7	**57.** 3	**58.** 8	**59.** 5	**60.** 1
61. 9	**62.** 0	**63.** 5	**64.** 6	**65.** 9
66. 6	**67.** 9	**68.** 3	**69.** 2	**70.** 5
71. 4	**72.** 4	**73.** 9	**74.** 2	**75.** 9
76. 0	**77.** 4	**78.** 9	**79.** 9	**80.** 1
81. 7	**82.** 1	**83.** 1	**84.** 6	**85.** 3
86. 8	**87.** 5	**88.** 8	**89.** 0	**90.** 5

PRACTICE EXERCISE 52

1. $8 \div 2 = 4$

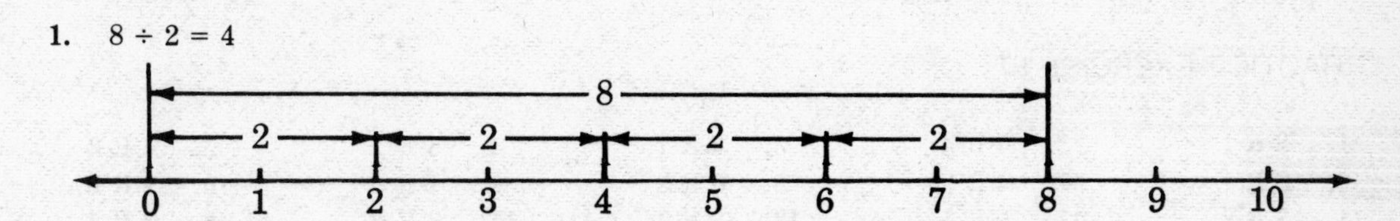

2. $10 \div 3 = 3\ R\ 1$

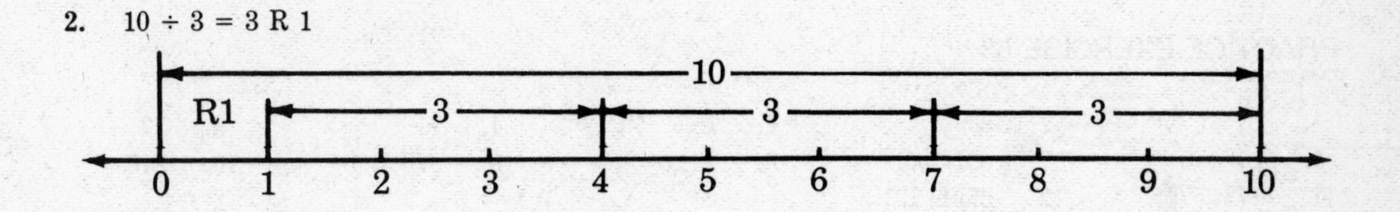

3. $16 \div 4 = 4$

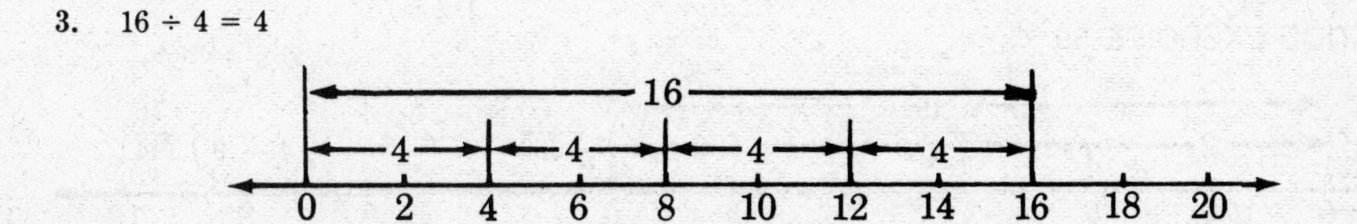

4. $11 \div 5 = 2\ R\ 1$

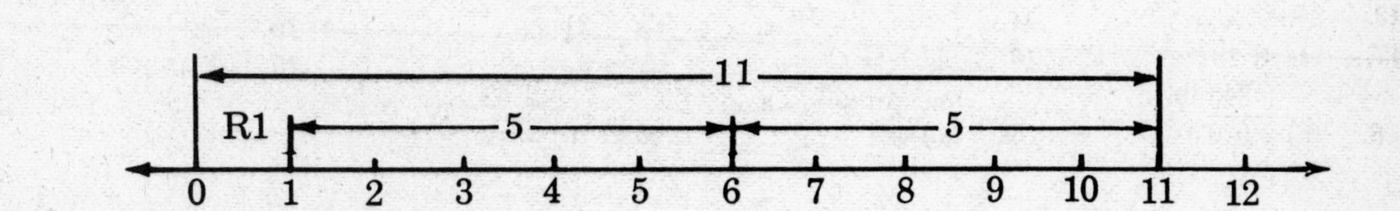

PRACTICE EXERCISE 53

1. 31	**2.** 21	**3.** 11	**4.** 43	**5.** 23
6. 12	**7.** 11	**8.** 31	**9.** 11	**10.** 12

PRACTICE EXERCISE 54

1. 223   2. 201   3. 4,321   4. 101   5. 3,112
6. 101   7. 421   8. 2,011   9. 13,101   10. 1,111
11. 10,012   12. 111   13. 1,012   14. 322   15. 434

PRACTICE EXERCISE 55

1. 61   2. 31   3. 72   4. 51   5. 81
6. 81   7. 31   8. 31   9. 64   10. 81

PRACTICE EXERCISE 56

1. 104   2. 40   3. 105   4. 106   5. 105
6. 90   7. 90   8. 102   9. 1,305   10. 1,031
11. 1,220   12. 1,060

PRACTICE EXERCISE 57

1. 8 R 1   2. 7 R 4   3. 7 R 1   4. 5 R 1   5. 4 R 3
6. 8 R 1   7. 4 R 3   8. 6 R 2   9. 7 R 5   10. 9 R 2
11. 7 R 3   12. 9 R 1   13. 8 R 3   14. 3 R 2   15. 3 R 5

PRACTICE EXERCISE 58

1. 13   2. 83   3. 19   4. 71   5. 77
6. 29 R 1   7. 651   8. 791   9. 679 R 1   10. 813
11. 778   12. 3,402 R 3

PRACTICE EXERCISE 59

1. $\frac{12}{6} = 2$ 2. $\begin{array}{r} 19\text{¢} \\ 3\overline{)\,57\text{¢}} \end{array}$ 3. $\begin{array}{r} 50 \text{ lb.} \\ 7\overline{)\,350} \end{array}$ 4. $\begin{array}{r} 16 \\ 9\overline{)\,144} \end{array}$

5. 52¢ ÷ 4 = 13¢   6. \$210 ÷ 3 = \$70   7. $\begin{array}{r} \$2400 \\ -\ 400 \\ \hline \$2000 \end{array}$ $\begin{array}{r} \$\ 250 \\ 8\overline{)\,\$2{,}000} \end{array}$

8. $\begin{array}{r} 345 \text{ lb.} \\ 6\overline{)\,2{,}070} \end{array}$ 9. $\begin{array}{r} 53 \text{ gal.} \\ 8\overline{)\,424} \end{array}$ 10. $\begin{array}{r} \$\ 52 \\ 6\overline{)\,\$312} \end{array}$

PRACTICE EXERCISE 60

1. 6   2. 13   3. 39   4. 46
5. 3   6. 3 R 6   7. 26   8. 1,142
9. 91   10. 88   11. 102   12. 72 R 18
13. 18 R 4   14. 57   15. 128 R 64   16. 3

PRACTICE EXERCISE 61

1. $\begin{array}{r} 27 \\ 8\,\overline{)\,216} \end{array}$

2. $\begin{array}{r} \$\ 623 \\ 12\,\overline{)\,\$7{,}476} \end{array}$

3. $\begin{array}{r} \$\ 47 \\ 24\,\overline{)\,\$1{,}128} \end{array}$

4. $\begin{array}{r} 63 \text{ lb.} \\ 108\,\overline{)\,6{,}804} \end{array}$

5. $\begin{array}{r} 12 \\ 16\,\overline{)\,192} \end{array}$

6. $\begin{array}{r} \$\ 165 \\ 47\,\overline{)\,\$7{,}755} \end{array}$

7. $\begin{array}{r} 1{,}372 \\ 12\,\overline{)\,16{,}464} \end{array}$

8. $\begin{array}{r} 35 \text{ gal.} \\ 12\,\overline{)\,420} \end{array}$

9. $\begin{array}{r} 8 \\ 54\,\overline{)\,432} \end{array}$

10. $\begin{array}{r} 37 \\ 375\,\overline{)\,13{,}875} \end{array}$

PRACTICE EXERCISE 62

1. $\dfrac{132 + 117 + 94 + 73}{4} = \dfrac{416}{4} = 104 \text{ lb.}$

2. a. $\dfrac{85\% + 94\% + 64\% + 88\% + 89\%}{5} = \dfrac{420}{5} = 84\%$

   b. 88%

   c. No

3. a. $\dfrac{13 + 11 + 5 + 11 + 10 + 9 + 4}{7} = \dfrac{63}{7} = 9 \text{ gal.}$

   b. 10 gal.

   c. Yes. 11 gal.

4. $\begin{array}{r} 232 \text{ lb.} \\ 11\,\overline{)\,2{,}552} \end{array}$

5. $\begin{array}{r} \$\ 73 \\ 5\,\overline{)\,\$365} \end{array}$

6. $\dfrac{63 + 61 + 57 + 54 + 43 + 40}{6} = \dfrac{318}{6} = 53$

PRACTICE EXERCISE 63

1. $43 \times ? = 1{,}462$
   $? = 34 \text{ ft.}$

2. $34 \times ? = 612$
   $? = 18 \text{ ft.}$

3. $12 \times ? = 108$
   $? = 9 \text{ in.}$

CROSS-NUMBER PUZZLE SOLUTION

Division of Whole Numbers

CROSS-NUMBER PUZZLE SOLUTION

All Operations

All Operations

(1) 3	(2) 7		(3) 5	6			(4) 9	(5) 8		(6) 6	(7) 3
(8) 1	4		4	■	(9) 9	(10) 2	■	1		(11) 9	6
	8		3	■	(12) 1	5	■	2		1	
(13) 1	3	2	■	(14) 1	3	7	(15) 2	■	(16) 1	2	(17) 1
9	■	■	(18) 7	8	■	■	(19) 1	(20) 8	■	■	2
	(21) 2	(22) 8	4	■	■	■	■	(23) 7	(24) 7	(25) 7	
	(26) 7	1	3	■	■	■	■	(27) 2	1	0	
(28) 3	■	■	(29) 4	(30) 2	■	■	(31) 5	0	■	■	(32) 6
(33) 5	(34) 9	2	■	(35) 7	(36) 1	(37) 2	1	■	(38) 1	(39) 2	3
	3		(40) 1	■	(41) 4	9	■	(42) 4		4	
(43) 3	2		6	■	(44) 4	3	■	3		(45) 5	(46) 6
(47) 6	0		(48) 9	0			(49) 3	9		(50) 8	5

With the completion of this puzzle you have finished the *four* fundamental operations for whole numbers. You are ready for Book 2, which introduces the fraction. Before you begin though, there are a few points that you should be aware of.

It takes time and a lot of your energy to succeed.

You will make errors as you study and do mathematics.

The number of your errors will decrease as you continue studying.

The idea that you must be perfect or that you must score a perfect paper can act like a brake to slow you down.

Although each pretest and posttest does not require a perfect paper, do try to achieve at least the satisfactory grade each time.

Hold It!

Are you developing your ability in learning to compute? Try this review exercise to see your progress in the four fundamental operations.

1. 103×35
2. $3{,}220 + 758 + 2{,}608 =$
3. $7{,}003 - 2{,}746$
4. $28 \overline{)392}$
5. $106 \overline{)25{,}122}$